BIOTECHNOLOGY BY OPEN LEARNING

Genome Management in Prokaryotes

PUBLISHED ON BEHALF OF :

Open universiteit and **University of Greenwich (formerly Thames Polytechnic)**

Valkenburgerweg 167
6401 DL Heerlen
Nederland

Avery Hill Road
Eltham, London SE9 2HB
United Kingdom

Butterworth-Heinemann Ltd
Linacre House, Jordan Hill, Oxford OX2 8DP

A member of the Reed Elsevier group

OXFORD LONDON BOSTON
MUNICH NEW DELHI SINGAPORE SYDNEY
TOKYO TORONTO WELLINGTON

First published 1993

British Library Cataloguing in Publication Data
A catalogue record for this book is
available from the British Library

Library of Congress Cataloguing in Publication Data
A catalogue record for this book is
available from the Library of Congress

ISBN 0 7506 0557 X

Composition by University of Greenwich
(formerly Thames Polytechnic)
Printed and Bound in Great Britain by
Thomson Litho, East Kilbride, Scotland

Genome
Management
in Prokaryotes

BOOKS IN THE BIOTOL SERIES

The Molecular Fabric of Cells
Infrastructure and Activities of Cells

Techniques used in Bioproduct Analysis
Analysis of Amino Acids, Proteins and Nucleic Acids
Analysis of Carbohydrates and Lipids

Principles of Cell Energetics
Energy Sources for Cells
Biosynthesis and the Integration of Cell Metabolism

Genome Management in Prokaryotes
Genome Management in Eukaryotes

Crop Physiology
Crop Productivity

Functional Physiology
Cellular Interactions and Immunobiology
Defence Mechanisms

Bioprocess Technology: Modelling and Transport Phenomena
Operational Modes of Bioreactors

In vitro Cultivation of Micro-organisms
In vitro Cultivation of Plant Cells
In vitro Cultivation of Animal Cells

Bioreactor Design and Product Yield
Product Recovery in Bioprocess Technology

Techniques for Engineering Genes
Strategies for Engineering Organisms

Principles of Enzymology for Technological Applications
Technological Applications of Biocatalysts
Technological Applications of Immunochemicals

Biotechnological Innovations in Health Care

Biotechnological Innovations in Crop Improvement
Biotechnological Innovations in Animal Productivity

Biotechnological Innovations in Energy and Environmental Management

Biotechnological Innovations in Chemical Synthesis

Biotechnological Innovations in Food Processing

Biotechnology Source Book: Safety, Good Practice and Regulatory Affairs

The Biotol Project

The BIOTOL team

OPEN UNIVERSITEIT, THE NETHERLANDS
Prof M. C. E. van Dam-Mieras
Prof W. H. de Jeu
Prof J. de Vries

UNIVERSITY OF GREENWICH (FORMERLY THAMES POLYTECHNIC), UK
Prof B. R. Currell
Dr J. W. James
Dr C. K. Leach
Mr R. A. Patmore

This series of books has been developed through a collaboration between the Open universiteit of the Netherlands and University of Greenwich (formerly Thames Polytechnic) to provide a whole library of advanced level flexible learning materials including books, computer and video programmes. The series will be of particular value to those working in the chemical, pharmaceutical, health care, food and drinks, agriculture, and environmental, manufacturing and service industries. These industries will be increasingly faced with training problems as the use of biologically based techniques replaces or enhances chemical ones or indeed allows the development of products previously impossible.

The BIOTOL books may be studied privately, but specifically they provide a cost-effective major resource for in-house company training and are the basis for a wider range of courses (open, distance or traditional) from universities which, with practical and tutorial support, lead to recognised qualifications. There is a developing network of institutions throughout Europe to offer tutorial and practical support and courses based on BIOTOL both for those newly entering the field of biotechnology and for graduates looking for more advanced training. BIOTOL is for any one wishing to know about and use the principles and techniques of modern biotechnology whether they are technicians needing further education, new graduates wishing to extend their knowledge, mature staff faced with changing work or a new career, managers unfamiliar with the new technology or those returning to work after a career break.

Our learning texts, written in an informal and friendly style, embody the best characteristics of both open and distance learning to provide a flexible resource for individuals, training organisations, polytechnics and universities, and professional bodies. The content of each book has been carefully worked out between teachers and industry to lead students through a programme of work so that they may achieve clearly stated learning objectives. There are activities and exercises throughout the books, and self assessment questions that allow students to check their own progress and receive any necessary remedial help.

The books, within the series, are modular allowing students to select their own entry point depending on their knowledge and previous experience. These texts therefore remove the necessity for students to attend institution based lectures at specific times and places, bringing a new freedom to study their chosen subject at the time they need and a pace and place to suit them. This same freedom is highly beneficial to industry since staff can receive training without spending significant periods away from the workplace attending lectures and courses, and without altering work patterns.

Contributors

AUTHORS

Dr J.J. Gaffney, Manchester Metropolitan University, Manchester, UK

Dr J.S. Gartland, Dundee Institute of Technology, Dundee, UK

Dr K.M.A. Gartland, Dundee Institute of Technology, Dundee, UK

Dr P. Hooley, University of Wolverhampton, Wolverhampton, UK

Dr S.H. Kirk, Nottingham Trent University, Nottingham, UK

Dr C.A. Smith, Manchester Metropolitan University, Manchester, UK

EDITOR

Dr K.M.A. Gartland, Dundee Institute of Technology, Dundee, UK

SCIENTIFIC AND COURSE ADVISORS

Prof M. C. E. van Dam-Mieras, Open universiteit, Heerlen, The Netherlands

Dr C. K. Leach, Leicester Polytechnic, Leicester, UK

ACKNOWLEDGEMENTS

Grateful thanks are extended, not only to the authors, editors and course advisors, but to all those who have contributed to the development and production of this book. They include Ms S. Connor, Ms H. Leather, Ms A. Liney, Ms J. Skelton, and Professor R. Spier.

The development of this BIOTOL text has been funded by **COMETT, The European Community Action Programme for Education and Training for Technology**. Additional support was received from the Open universiteit of The Netherlands and by University of Greenwich (formerly Thames Polytechnic).

Contents

How to use an open learning text

An open learning text presents to you a very carefully thought out programme of study to achieve stated learning objectives, just as a lecturer does. Rather than just listening to a lecture once, and trying to make notes at the same time, you can with a BIOTOL text study it at your own pace, go back over bits you are unsure about and study wherever you choose. Of great importance are the self assessment questions (SAQs) which challenge your understanding and progress and the responses which provide some help if you have had difficulty. These SAQs are carefully thought out to check that you are indeed achieving the set objectives and therefore are a very important part of your study. Every so often in the text you will find the symbol Π, our open door to learning, which indicates an activity for you to do. You will probably find that this participation is a great help to learning so it is important not to skip it.

Whilst you can, as an open learner, study where and when you want, do try to find a place where you can work without disturbance. Most students aim to study a certain number of hours each day or each weekend. If you decide to study for several hours at once, take short breaks of five to ten minutes regularly as it helps to maintain a higher level of overall concentration.

Before you begin a detailed reading of the text, familiarise yourself with the general layout of the material. Have a look at the contents of the various chapters and flip through the pages to get a general impression of the way the subject is dealt with. Forget the old taboo of not writing in books. There is room for your comments, notes and answers; use it and make the book your own personal study record for future revision and reference.

At intervals you will find a summary and list of objectives. The summary will emphasise the important points covered by the material that you have read and the objectives will give you a check list of the things you should then be able to achieve. There are notes in the left hand margin, to help orientate you and emphasise new and important messages.

BIOTOL will be used by universities, polytechnics and colleges as well as industrial training organisations and professional bodies. The texts will form a basis for flexible courses of all types leading to certificates, diplomas and degrees often through credit accumulation and transfer arrangements. In future there will be additional resources available including videos and computer based training programmes.

Preface

The study of microbial and molecular genetics holds a key position in contemporary biology and biotechnology. It was through the study of inheritance amongst bacteria that Avery and his co-workers first established that DNA was the chemical which encoded genetic information. Subsequent studies with these organisms facilitated the many advances that have been made in our understanding of gene organisation, expression and regulation. These advances include elucidation of the major processes of transcription and translation, the discovery of operons and regulatory genes (sequences) and, more recently, the discovery of restriction endonucleases. The study of microbial genetics, particularly the genetics of bacteria, and their associated viruses, has been monumental in its contribution to our understanding of the genetic and molecular processes of organisms in general.

The importance of bacterial molecular genetics is not, however, confined to its contribution to biological knowledge. This knowledge has had enormous inpact in such diverse and essential areas as healthcare, agriculture food processing and environmental management. The discovery of restriction endonucleases has been particularly significant. We are still assimilating the potential uses of the knowledge and techniques that have arisen from the study of bacterial molecular genetics. It is, however, widely recognised that these practical applications extend far beyond simply manipulating the genetics and phenotypic characteristics of micro-organisms. The study of the molecular genetics of prokaryotes is, therefore, of fundamental academic and practical importance.

This book is designed to provide readers with a thorough understanding of the key features and properties of the genetic exchange systems that occur in prokaryotic cells and explains how genes in these systems are expressed and regulated. It also aims to provide readers with an understanding of a range of techniques that are used in microbial genetics. This knowledge provides the underpinning framework within which the reader may go on to learn how to clone, analyse and manipulate specific genes or sets of genes and to engineer new organisms.

The authors have taken great care in selecting from the mountain of information that has emerged about the genetic systems of prokaryotes to provide a coherent and well-balanced story which covers all of the key issues. In doing so they have managed to avoid the dangers of producing a tedious encyclopedia and to produce a text enjoyable in terms of pace and space!

The text has been written on the assumption that the reader has some previous knowledge of cell biology especially about the structure and major chemical constituents of prokaryotic cells. The authors have, however, provided many helpful reminders. The text begins by describing the structure and properties of DNA and goes on to discuss how genes may be altered by mutagenesis. In Chapters 3 and 4, the authors describe the processes by which genes may be exchanged and re-arranged in prokaryotes. This description includes transformation, conjugation, transduction and transposition. The remaining chapters deal with the issues of gene expression and the mechanisms by which gene expression may be regulated. Thus Chapters 5 and 6 follow the temporal sequence found in cells by describing the processes of transcription and the translation of transcription products into proteins. In Chapters 7 and 8, the regulation of gene expression is described using the regulation of important operons to illustrate the range of mechanisms used by bacteria.

This book is partnered by another BIOTOL text, 'Genome Management in Eukaryotes', which builds on the material covered in the text by examining the special features of eukaryotic molecular genetics. Together these texts provide a substantive study which will enable readers to develop knowledge and expertise in recombinant DNA technology and genetic engineering covered in the BIOTOL texts, 'Techniques for Engineering Genes' and 'Strategies for Engineering Organisms'.

Scientific and Course Advisors: Professor M.C.E. van Dam-Mieras
Dr. C.K. Leach

DNA structure and replication in prokaryotes and viruses

DNA structure and replication in prokaryotes and viruses

1.1 Introduction

A great deal of what has come to be called biotechnology is concerned with the experimental and commercial uses of molecules called nucleic acids, which were first discovered in the nineteenth century. Biotechnology, however, has been dependent upon basic research into the structure and function of these compounds. In the twentieth century brilliant investigations by Griffiths (1928), Avery *et al* (1944) demonstrated that DNA carries specific genetic information and Watson and Crick (1953) elucidated its structure. These advances laid foundations in molecular biology and stimulated many studies in nucleic acid biochemistry.

In this chapter we are going to examine the structure and properties of DNA and explain how these molecules are replicated in prokaryotes (bacteria) and viruses. We will first briefly describe how these molecules were discovered and how they were identified as the molecules which carry genetic information. We will then describe the structure of these molecules in some detail together with their interactions with other molecules. This will provide the basis for the description of how nucleic acids are packaged and replicated within cells. This knowledge underpins the material covered within the remainder of the text.

1.1.1 The discovery of nucleic acids

In 1868 Miescher isolated a phosphorus-rich material from a crude preparation of nuclei obtained from white blood cells. He called this substance nuclein. It is now known that nuclein is a complex of nucleic acid and protein, ie it is a nucleoprotein. Similar material was soon isolated from a number of tissues and it gradually became apparent that nucleoproteins are constituents of all cells. A protein-free preparation was eventually isolated by Altman in 1898 who coined the term nucleic acid for the material. The specific nucleic acid obtained from nuclei is deoxyribonucleic acid (DNA).

Hydrolysis of DNA and subsequent analysis of the products of hydrolysis showed it was composed of the purine (Pu) bases adenine (A) and guanine (G) and the pyrimidine (Py) bases cytosine (C) and thymine (T), a deoxypentose sugar (now known to be deoxyribose, dRib) and phosphate (Figure 1.1).

However, a nucleic acid originally obtained from yeast differed in that it contained uracil (U) not thymine and a pentose (ribose, Rib) rather than deoxypentose sugar (Figure 1.1). This nucleic acid is ribonucleic acid (RNA). It resembled a nucleic acid preparation from a plant and so for a number of years it was thought that plants contained RNA and animals DNA. By the mid-1920s it was realised that plant, animal and bacterial cells contain both DNA and RNA. It is worth noting at this stage that viruses, which lack a cellular structure, contain DNA or RNA but not both (see later).

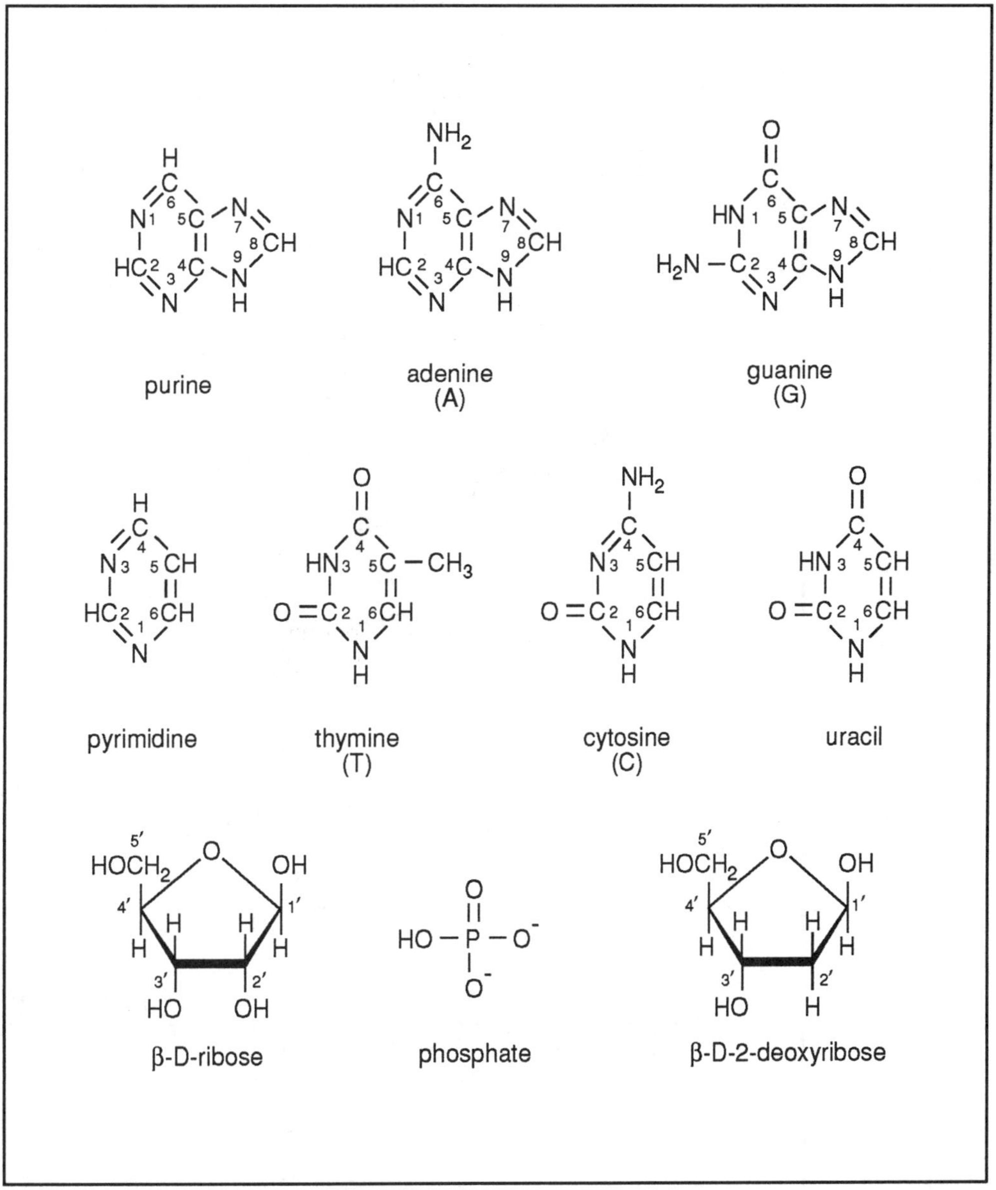

Figure 1.1 Components of nucleic acids. Note the system used to number the atoms present in the purine and pyrmidine rings. Numbers used to denote the carbon atoms in the two sugars are designated with a ' to distinguish them from those in the nitrogenous bases.

1.1.2 Nucleotides and nucleosides

A series of investigations which began as early as 1909 by Levene and Jacobs and culminated in 1951 in the work of Todd, showed that nucleic acids were composed of nucleotides (Figure 1.2).

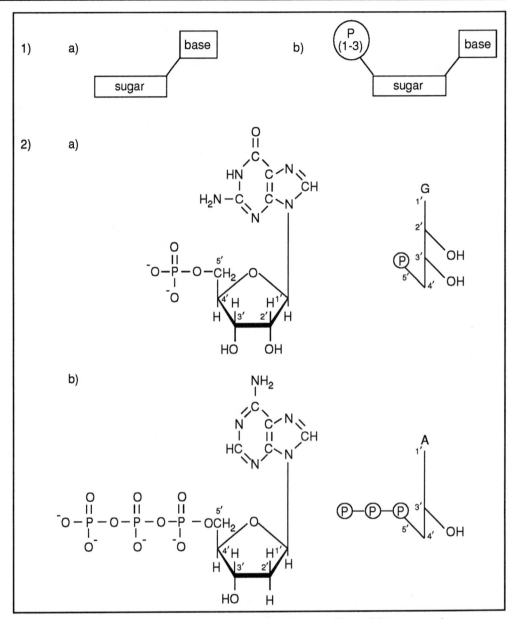

Figure 1.2 1) Outline of the structures of a) nucleosides and b) nucleotides and 2) two ways of representing nucleotides using a) guanosine 5'-monophosphate and b) deoxyadenosine 5'-triphosphate as examples. The positions in the sugars are indicated by the prime (') to distinguish them from similarly numbered base atoms. Note the stylised drawing of these nucleotides on the right of the figure. In these the sugar residues are shown simply as lines and the bases are shown by a letter.

nucleotide

nucleoside

Each nucleotide is a combination of base-sugar-phosphate(s). A base-sugar only combination is a nucleoside. DNA is composed of deoxyribonucleotides while RNA is formed of ribonucleotides. The nomenclature of deoxyribonucleosides and deoxyribonucleotides is given in Table 1.1.

Base	Deoxyribonucleoside	Deoxyribonucleotide (eg 5′- monophosphate)[1]
adenine	deoxyadenosine	deoxyadenosine 5′ - phosphate (dAMP)[2][3] (or deoxyadenylate)[4]
guanine	deoxyguanosine	deoxyguanosine 5′ - phosphate (dGMP) (or deoxyguanylate)[4]
cytosine	deoxcytidine	deoxycytidine 5′-phosphate (dCMP) (or deoxycytidylate)[4]
thymine	deoxythymidine [5]	deoxythymidine 5′ - phosphate (dTMP)[5] (or deoxythymidylate)[4]

Table 1.1 Nomenclature of deoxyribonucleosides and deoxyribonucleotides. [1] the single phosphate is carried on C-5′ of the deoxyribose; [2] the d denotes the presence of deoxyribose; [3] other esters must be specified eg 3′-dATP means a triphosphate group is carried on C-3′; [4] these compounds may also be named as the corresponding acids which are ionised at pH7; [5] sometimes called (confusingly) thymidine and thymidine 5′-phosphate (dTMP) respectively.

SAQ 1.1

Using Table 1.1 as a model, give an appropriate nomenclature for ribonucleosides and ribonucleotides.

nucleic acid

phosphodiester bonds

In nucleic acids, nucleotides are joined together by phosphodiester bonds (Figure 1.3) to form high molecular weight polymers. However, in 1951 the overall structure of DNA was not known nor, indeed, was its biological role fully appreciated.

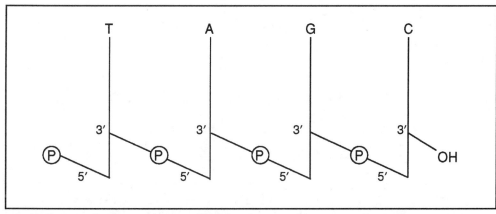

Figure 1.3 Phosphodiester bonds in DNA. The molecule is drawn with a 5′ → 3′ polarity. The sequence shown may be abbreviated to pdTpdApdGpdC or pdTAGC, where p denotes a phosphate group and d recognises the presence of deoxyribose. Note the importance of indicating the presence or absence of terminal phosphate groups. In this figure we have used the stylised representation described in Figure 1.2. Note that the molecule depicted in this figure is said to have 5′ → 3′ polarity.

1.2 Bacterial transformation and DNA

S and R
colonies

The bacterium *Streptococcus pneumoniae* can occur in a number of identifiable strains eg type I, II, III, each of which can form smooth(S) or rough(R) colonies when grown on agar plates. The S colonies result from the presence of a genetically determined carbohydrate-rich capsule which surrounds each bacterial cell - hence the glistening appearance of the colonies. Different strains of S bacteria possess chemically different capsules. Bacteria of R colonies lack capsules. The capsules function to protect the bacteria from the host's immune system, and the S forms of all strains cause a fatal pneumonia (ie are virulent) when injected into mice. R forms, however, are non-pathogenic. In 1928, Griffiths showed that a non-pathogenic R type could be transformed into a pathogenic S type by mixing live R type with S type cells which had been heat killed. Indeed, a cell-free extract of the S type III strain mixed with live R type II cells *in vitro* was virulent when injected into a mouse. Furthermore, the bacteria subsequently isolated from the mouse serum had capsules associated with type III *Streptococcus*! Growth of these bacteria over a number of generations showed the ability to form the capsule was now innate. Thus the type II cells had somehow acquired new genetic information from the cell-free extract of the type III strain.

transformation

A series of marvellously detailed investigations by Avery and co- workers, reported in 1944, showed that transformation in *Streptococcus* was mediated by DNA. Therefore the genetic information needed to specify capsule formation was carried by DNA. Previous to these studies, DNA had been ascribed a mere scaffold-type role in the nucleus with genetic information being carried by proteinaceous genes. However, Avery's work was slow to gain acceptance. The structure of DNA was not known, therefore the manner in which it could carry hereditary information and transmit it to subsequent generations during reproduction could not be envisaged.

The answers to these problems were supplied by the elucidation of the double helical structure of DNA by Watson and Crick.

1.3 The double helical structure of DNA

Watson and Crick began to investigate the structure of DNA in 1951. Their approach was to attempt to build a molecular model of DNA which was consistent with the known data about its structure obtained by other investigators.

x-ray
diffraction
studies

X-ray diffraction studies on fibres of DNA by Astbury in the 1930s had indicated the molecules were long and thin with the bases stacked at right angles to the long axis. Extensive X-ray diffraction studies by Wilkins and Franklin in the early 1950s indicated that DNA could exist in one of at least two possible conformations. Diffraction patterns obtained when fibres of DNA were X-rayed in an atmosphere of low relative humidity arose from an A-form of DNA. This changed to a B-form at high relative humidity. Both of the resulting diffraction patterns clearly resulted from helical molecules. However, B-DNA give a simpler, more easily interpreted pattern, with strong 0.34 and 3.4 nm spacings which corresponded to the distances between adjacent bases and the length of a complete turn of the helix (ie the pitch) respectively. The pattern also indicated that the molecule was asymmetric along the long axis: a feature which suggested a double-stranded helix with the strands running in opposite directions.

Chargaff's rule A further clue was provided from studies by Chargaff on the composition of DNA from a number of sources. These indicated that, within experimental limits, the molar amount of A was the same as that of T and that G equalled C (Table 1.2).

Organism[1]	A	G	C	5-MeC[2]	T	A/T	G/C
Vaccinia virus	29.5	20.6	20.0	-	29.9		
E. coli	25.7	24.2	24.6	-	25.5		
Yeast	31.7	18.3	17.4	-	32.6		
Chlorella vulgaris	20.2	30.0	26.4	3.45	19.8		
Broad bean	29.7	20.6	14.9	5.2	29.6		
Herring	27.8	22.2	20.7	1.9	27.5		
Frog	26.3	23.5	21.8	2.0	26.4		
Chicken	28.9	23.7	20.3	0.91	26.2		
Human	30.3	19.5	19.9	-	30.3		

Table 1.2 Shows the relative amounts of bases in the DNA from a number of species (see also SAQ 1.2).
[1] Data from a variety of sources
[2] 5-MeC, 5-methylcytosine is a modified form of cytosine (see Figures 1.1 and 1.7) which is common in organisms; it constitutes part of the cytosine component.

basepairing Watson and Crick observed that models of A and T and G and C could be arranged to give A-T and G-C basepairs (bp) which had similar overall shapes and sizes (Figure 1.4).

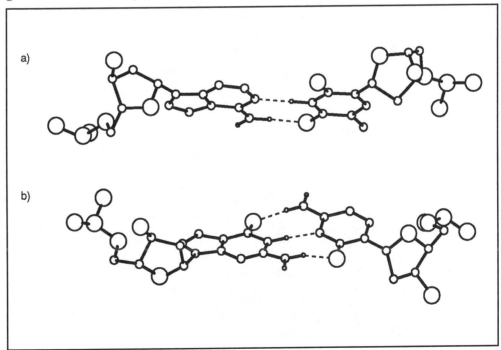

Figure 1.4 Molecular models of a) A-T and b) G-C basepairs. Only the hydrogens involved in hydrogen bonding (dotted lines) between the bases are shown. You might like to combine the information in this figure with that in Figure 1.1. Identify the various atoms in the bases that are illustrated.

The basepairs shown in Figure 1.4 are more commonly illustrated as shown below:

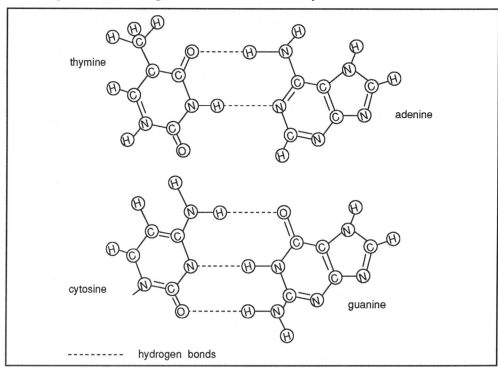

Figure 1.5 Common representation of hydrogen bonding in basepairing in DNA.

SAQ 1.2

1) Does the data in Table 1.2 support Chargaff's ratios?

2) From Chargaff's rules what can be deduced about the relative amounts of Pu and Py bases?

1.3.1 Watson and Crick structure of B-DNA

In 1953 Watson and Crick finally managed to build a molecular model of B-DNA which satisfied all that was then known of its structure. This model is the now famous double helix. The structure of B-DNA consists of two polydeoxyribonucleotide chains. Each is a right-handed helix which twists about a common imaginary central axis to give a double helix of about 2.0 nm diameter (Figure 1.6). Figure 1.6 shows several types of models of DNA. The first half of the figure shows a common representation of B-DNA in which the sugar-phosphate chains are shown as ribbons with the bases extending into the inner parts of the helix. Note that G always pairs with C and T with A. Note that the hydrogen bonds between the basepairs are also included. In the second half of the figure, stylised molecular models are represented in which the deoxyribose rings can be seen and the basepairs are shown as flat plates between the sugar-phosphate chains. The first half of this figure is a model of the A-form of DNA, whilst the B-form is shown in the second half.

major and minor grooves

The sugar-phosphate groups are on the outside forming two continuous backbones, with the bases projecting into the centre of the molecule, rather like the rungs of a ladder. The surface of the molecule has two continuous grooves: a wider, major and a narrower, minor groove. The chains are anti-parallel, that is one helix runs in the $5' \rightarrow 3'$, the other in the $3' \rightarrow 5'$ direction.

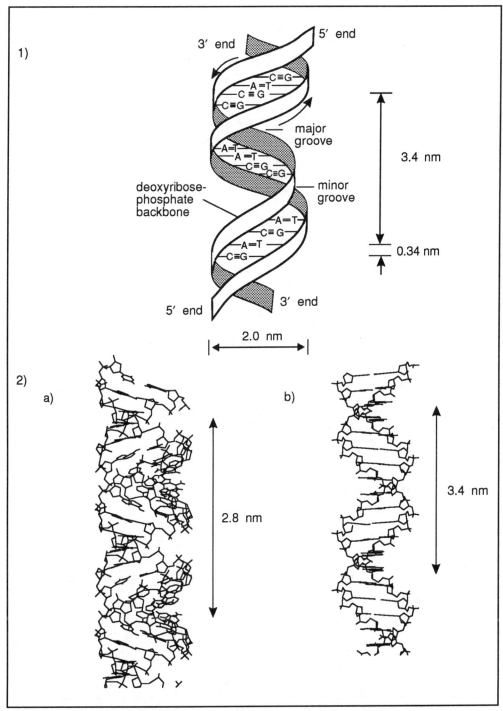

Figure 1.6 1) Schematic outline of the structure of B-DNA. 2) a) A-DNA and b) B-DNA. Note both are right handed helices. A-DNA has a rotation of 33° per residue and the pitch per spiral is about 2.8 nm. B-DNA has a rotation of about 36° per residue and a pitch of about 3.4 nm.

Π Look back at Figure 1.3 to remind yourself of the polarity of DNA strands. Then
    draw the strand which would basepair with the strand shown in Figure 1.3.

You should have drawn the strand:

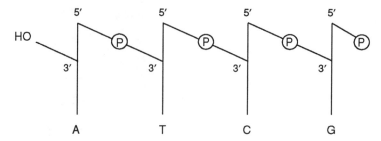

Note that this would basepair with the strand shown in Figure 1.3. We could represent
the double-stranded molecule that would be formed in the following way:

$$
\begin{array}{cccccc}
3' & A & T & C & G & 5' \\
   & || & || & ||| & ||| & \\
5' & T & A & G & C & 3'
\end{array}
$$

Now let us return to the three dimensional structure of DNA.

A consequence of the anti-parallel nature of the strands is that the double helical
structure does not begin to repeat itself until the strands have progressed through 360°.

In B-DNA, ten basepairs occupy one complete turn of the double helix which is 3.4 nm
long. Hence each basepair is rotated by 36° relative to, and is 0.34 nm distant from, the
next.

Π What would be the structural consequences if the deoxyribonucleotide chains of
    DNA were parallel rather than anti- parallel? Assume the same dimensions for a
    basepair and turn of helix.

If the strands were parallel, the overall structure would repeat itself after 180°, not 360°
as with antiparallel chains. Thus each base would be 0.34 nm from its neighbours but
only rotated by 18° relative to adjacent bases (not 36°). When Watson and Crick tried to
build models with parallel chains, they found it very difficult because 18° did not leave
sufficient room to fit in the bases.

The basepairs are arranged at roughly right angles to the sugar-phosphate backbones
with hydrogen bonding between complementary basepairs holding the two helices
together. A always pairs with T and G with C otherwise the diameter of helix would
not be regular. This feature explains Chargaff's ratios. The G-C pair has three hydrogen
bonds, the A-T pair two as shown in Figures 1.4 and 1.6. It follows from the
complementary pairing that the order of bases in one strand determines the order in the
other. There is, however, no restriction on the order of bases within a strand. Note that
DNA can exist in two forms A-DNA and B-DNA. The structure shown in Figure 1.6a is
the B-DNA form. Molecular models of both A and B forms are shown in Figure 1.6b. It
is the B-form which is normally encountered in nature.

SAQ 1.3	1) If one strand of a DNA molecule contains 18% A, what can be deduced about a) the composition of that strand and b) the composition of the complementary strand? 2) If one strand of a DNA molecule has the base sequence: 5′ ... AATTTGCCGGATAGGCCCAT ... 3′ what is the primary structure of its complementary strand?

1.3.2 Genetic implications of the Watson-Crick model

The double helical structure of DNA proposed by Watson and Crick had obvious implications with regard to its role as the carrier of genetic information. The sequence of bases is the primary structure of DNA. By convention, primary structures of nucleic acids are written with the 5′ end to the left and the 3′ to the right.

∏ How would the primary structures of DNA from closely related organisms be expected to vary when compared with those of distantly related organisms?

You should have argued that closely related organisms would have a large amount of genetic information in common and would, therefore, have DNA with very similar sequences (ie primary structures). The more distantly related the organisms, the less common the primary structures of their DNAs.

Both types of nucleic acids (DNA and RNA) are concerned with storing and expressing biological, that is genetic, information. DNA is the repository of genetic information. In pro- and eukaryotes RNA is intimately concerned with the expression of this information. All diploid cells of a particular organism will have identical DNA, which will have a different primary structure to the DNA of other organisms.

Viruses are unusual in that their nucleic acid may be DNA or RNA either of which can be double or single-stranded. Linear or circular molecules may be present depending upon the type of virus (Table 1.3).

Virus*	Nucleic acid
φX-174, hepatitis	single-stranded, circular DNA
pneumonia, bacteriophage T2	double-stranded, linear DNA
papilloma	double-stranded, circular DNA
tobacco mosaic virus (TMV)	single-stranded, circular RNA
infantile gastroenteritis	double-stranded, linear RNA

Table 1.3 Viral nucleic acids. *Viruses are often named according to the disease they cause or are associated with.

| SAQ 1.4 | The DNA of the virus φX-174 is a single-stranded molecule containing 24.6% A. What can be deduced about the relative amounts of the other bases? |

1.3.3 Variations in the secondary structure of DNA

DNA secondary structure

The double helical structure of B-DNA is an example of a secondary structure of a nucleic acid. DNA shows other secondary structures (see below) and, indeed, is coiled and folded into even higher-ordered structures.

X-ray crystallography of fibres of DNA by Franklin had indicated that DNA could occur in at least two possible conformations: A- and B-DNA. However, X-ray analysis of fibres is a relatively crude method and gives only an overall impression of structure. Detailed structure ie at the molecular level requires X-ray analysis of crystals. In the late 1970s, crystallisation of synthetic oligodeoxyribonucleotides was accomplished. X-ray crystallography of the crystal form of:

```
5'   C G C G A A T T C G C G    3'
     | | | | | | | | | | | |
3'   G C G C T T A A G C G C    5'
```

showed it to have the expected B-DNA structure, while that of:

```
5'    G G T A T A C C    3'
      | | | | | | | |
3'    C C A T A T G G    5'
```

showed it to have an A-type structure. (You should re-read the legend to Figure 1.5 to refresh your memory of some of these differences). Nevertheless, in both cases individual nucleotide residues departed significantly from the 'average' specifications suggested by the Watson and Crick model. For example, the helical twist in the B-DNA crystal varied between 28° - 42° compared with the value of 36° in the B-model. Further, the variations occurred in a manner which depended upon the primary structure, that is the base sequence.

Z-DNA

In contrast to the above, X-ray crystallography of crystals of the oligodeoxyribonucleotides:

```
CGCGCG  and  CGCG
| | | | | |      | | | |
GCGCGC       GCGC
```

revealed a very different structure: a left-handed, thinner double helix. In this form, a line linking adjacent phosphate groups follows a zig-zag pathway. Hence this structural from is called Z-DNA. In Z-DNA, the helices are left-handed and there is a rotation of -30°. The pitch of the helix is 4.5 nm per turn and has a diameter of 1.84 nm (compared with 2.55 nm for A-DNA and 2.37 nm for B-DNA.

Z-DNA is favoured by alternating Pu-Py sequences, and high ambient salt concentrations which stabilise the structure by reducing the repulsive electrostatic forces between the phosphate groups on opposite strands (Z-DNA is narrower than both A- and B-DNA, see Table 4.1). However, Z-DNA can occur at physiological ionic strengths if a significant proportion of the cytosine residues are methylated to form 5-methylcytosine (Figure 1.7), a modification which is common in many organisms

(Table 1.2). Z-DNA has been shown to be present in *Escherichia coli* cells, although its biological function is not known.

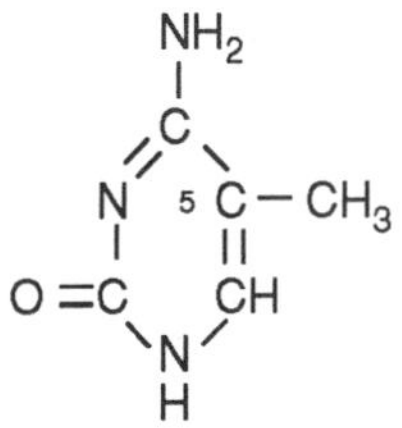

Figure 1.7 5-Methylcytosine.

∏ Suggest a plausible reason why methylation of cytosine (to 5-methylcytosine) may promote the formation of Z-DNA. Hint: think about hydrophobicity.

The hydrophobic nature of the methyl group means that in an aqueous environment the bases pack more tightly together giving the thinner double helix of Z-DNA.

	A	B	Z
handedness	?	?	?
number of bases/turn	11	10	12
rotation/residue (°)	?	?	?
rise/residue (nm)	0.255	0.34	0.37
pitch (nm)			
diameter (nm)	2.55	2.37	1.84

Table 1.4 A summary of the properties of A-, B- and Z-DNA (see SAQ 1.5).

SAQ 1.5

Table 1.4 summarises some of the properties of A-, B- and Z-DNA. It is, however, incomplete.

Complete Table 1.4 to summarise some of the structural differences between A-, B- and Z-DNA.

Cruciforms

cruciforms

inverted repeat

Cruciforms (cross-shaped structures) are a secondary structure of DNA which occur when the sequence of bases on one strand is followed by the same sequence in reverse order on the opposite strand. This arrangement is called an inverted repeat. An inverted repeat must have a corresponding inverted repeat on the complementary strands in order for a cruciform to be produced. (Figure 1.8a). If the DNA is denatured, that is its secondary (double helical) structure disrupted so that the interchain hydrogen bonds are broken, a hairpin loop can form because of intrastrand hydrogen bonding, giving a cruciform structure (Figure 1.8b).

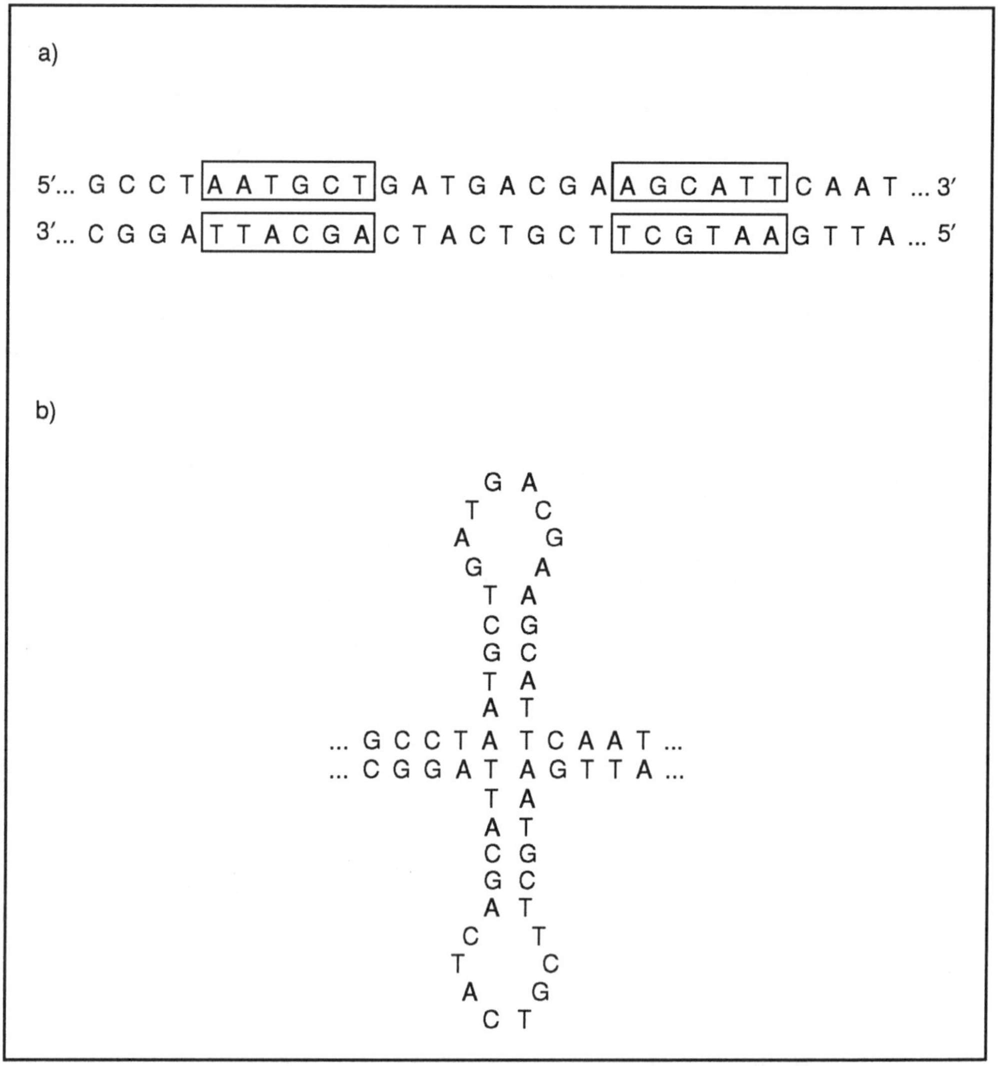

a)

5'... G C C T [A A T G C T] G A T G A C G A [A G C A T T] C A A T ... 3'
3'... C G G A [T T A C G A] C T A C T G C T [T C G T A A] G T T A ... 5'

b)

```
                        G  A
                     T        C
                   A            G
                   G            A
                      T      A
                      C      G
                      G      C
                      T      A
                      A      T
              ... G C C T A   T C A A T ...
              ... C G G A T   A G T T A ...
                      T      A
                      A      T
                      C      G
                      G      C
                      A      T
                   C            T
                   T            C
                   A            G
                      C  T
```

Figure 1.8 a) Inverted repeats (boxed) in a sequence of DNA. b) Cruciform structure resulting from intrastrand basepairing of the inverted repeats.

SAQ 1.6

1) Will the following deoxyribonucleotide sequence be capable of forming a cruciform?

        T G A C C G A A T T C C T C G G T C G
        A C T G G C T T A A G G A G C C A G C

2) If this is the case, which would be the more stable structure: the cruciform, or the double helix?

Cruciforms can only form when denaturation, that is unwinding of the DNA double helix, occurs. Hence any stress which promotes denaturation will favour their formation. It has been suggested that cruciforms may form recognition sites for the specific binding of proteins to DNA.

1.3.4 DNA and chromosomes

chromosomes   The B-structure of DNA first proposed by Watson and Crick has been modified and refined over the years since 1953. It does, however, seem to be the case that B-DNA is the type most prevalent in cells. Eukaryotic chromosomes each contain a single linear double-stranded DNA molecule. Prokaryotes, eg bacteria like *E. coli*, have chromosomes which contain a single circular double-stranded DNA molecule (Figure 1.9).

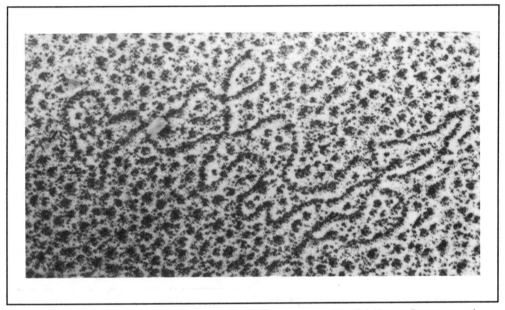

Figure 1.9 Electron micrograph of a circular DNA molecule. Courtesy of Dr. P A Kumar, Department of Biological Sciences, Manchester Polytechnic.

1.3.5 Supercoiling of DNA

supercoiled DNA   Circular double-stranded DNA occurs naturally in a tertiary structure called supercoiled DNA (synonyms superhelical and supertwisted DNA). Circular DNA which is not supercoiled is called relaxed DNA. Relaxed DNA, unlike supercoiled, will lie in a plane. If the strands of the molecule are cleaved and slightly unwound before resealing, then rather than existing as a planer molecule with an unwound loop, the double helix becomes distorted to form the more stable negatively supercoiled DNA (Figure 1.10). Negative means a left handed form as shown in the figure.

If the strands were cleaved and the helix wound slightly before resealing, then the double helix becomes distorted to form a positive supercoiled DNA. Positive means a right handed form. The induction of supercoiling is catalysed by an enzyme called DNA-gyrase.

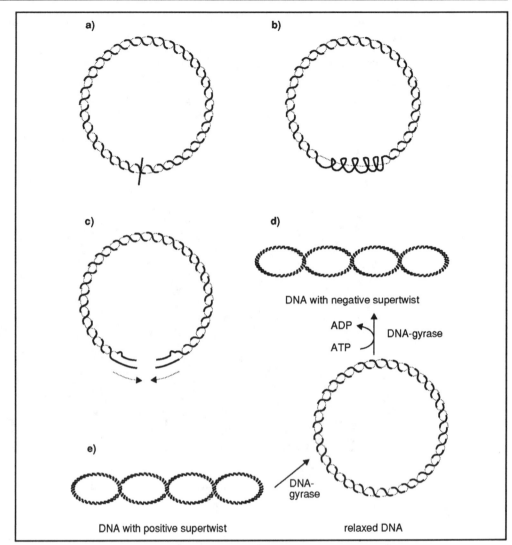

Figure 1.10 Supercoiling of DNA. a) If both strands of a supercoiled DNA molecule are broken and b) the strands slightly unwound and then c) resealed, the molecule will d) assume the more stable negative supercoiled conformation rather than a conformation with an open loop. e) shows a positive supercoiled conformation (see text). Redrawn from Wang J C (1982) Sci Amer 247(1), 84-95.

topoisomerases

gyrase

In bacterial cells, supercoiling of DNA is catalysed by topoisomerases. One particular type of topoisomerase called gyrase, which has only been found in bacteria, uses the free energy of hydrolysis of ATP to promote negative supercoiling.

$$\text{relaxed DNA} \quad \xrightarrow[\text{ATP}]{\text{gyrase}} \quad \text{negatively supercoiled DNA}$$

Supercoiled DNA is more compact than the relaxed form and can therefore be packaged in the restricted confines of a bacterial cell more readily (see later). Negative supercoiling causes unwinding of the double helix. Hence supercoiled DNA is primed for a number of biological processes which depend upon the unwinding of the double helix, such as replication (see later).

SAQ 1.7

Fill in the missing words or numbers using words or numbers selected from the list below:

DNA is composed of the sugar [   ], the bases [   ], [   ], [   ] and [   ] and [   ] groups. The bases of nucleic acids are classified into one of two groups, either [   ] or [   ]. A combination of base and sugar form a [   ]; a base-sugar-phosphate is a [   ]. Nucleic acids consist of [   ] linked by [   ] bonds.

Watson and Crick proposed a structure for DNA based on the [   ] data of Franklin and Wilkins. Their proposal was that DNA was [   ] with a distance between bases of [   ] nm and a pitch of [   ] nm. The proposed structure had an approximate diameter of [   ] nm. The two strands are [   ], one strand running in the [   ] direction and the other [   ].

DNA can assume one of a number of secondary structures. The most common is [   ] but other conformations such as [   ] and [   ] exist, [   ] is favoured at low relative humidity. [   ] is formed at high ambient salt concentrations and favoured by alternating [   ] sequences.

Word list: double helical, pyrimidines, A-DNA (two occurrences), deoxyribose, purines, nucleotides, adenine, nucleotide, Z-DNA (two occurrences), cytosine, 0.34, Pu-Py, thymine, 3.4, X-ray diffraction, 3′ → 5′, nucleoside, antiparallel, phosphodiester, 5′ → 3′, 2.0, phosphate, guanine, B-DNA.

1.3.6 RNA and helices

RNA

RNA molecules are generally single-stranded, although some double-stranded forms do occur in some viruses (Table 1.3). However, even single-stranded RNA molecules can form double helices in regions where the strand doubles back along itself (Figure 1.11).

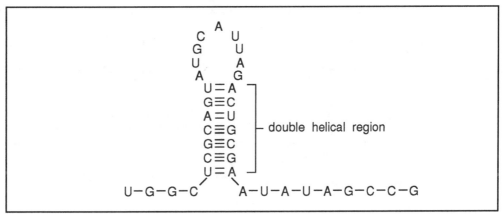

Figure 1.11 Double-stranded region in RNA. Note that in RNA, A pairs with U and contrasts with the pairing of A with T in DNA.

These double helical regions are stabilised by hydrogen bonds between complementary basepairs: the usual G-C pair and an A-U pair, which is also the basepair in 'genuine' double-stranded RNA molecules. Double helical RNA adopts an A-configuration, since the 2′-OH groups of the ribose residues prevent the formation of B-type structures.

heteroduplexes

Hetroduplexes, in which one strand of the double helix is DNA and the other RNA, can also form. These arise during some natural processes eg DNA replication. They may also be made artificially in the laboratory and are of importance to some biotechnological manipulations.

∏ Would you expect a heteroduplex, ie a DNA-RNA hybrid molecule, to adopt an A- or B-type structure?

You should have anticipated that it would be an A-type structure because the 2′-OH groups of the RNA strand will sterically prevent the formation of a B-type structure.

1.4 Interactions of DNA with proteins

It has been recognised for some time that the major and minor grooves of DNA (see Figure 1.5) could accommodate α-helix and β- sheet structures of proteins respectively. In recent years studies on specific interactions between DNA and proteins have been greatly advance by X-ray crystallography of a number of specific DNA-protein complexes, and by the use of computer-based molecular modelling systems. Such studies have determined a number of ways in which the association between DNA and DNA-binding proteins is stabilised. Broadly, the forces which stabilise these associations are the weak interactions which contribute to mutual binding in other biochemical situations, eg enzyme-substrate and hormone-receptor interactions, that is: hydrogen bonding, salt links, van der Waals and hydrophobic interactions.

General and specific recognition

general and
specific
recognition

The sugar-phosphate backbones of DNA are characterised by the repetition of many negative charges. Positively charged side chains of lysine and arginine residues in proteins may form ionic bonds with the phosphates groups. The association between protein and DNA stabilised in this manner would, however, be essentially independent of the base sequence of the DNA. In contrast, bases of DNA which line the major and minor grooves form patterns of chemical groups which can potentially form hydrogen bonds with proteins (see SAQ 1.8). This would, of course, allow proteins to recognise and interact with specific DNA sequences as long as more than one hydrogen bond was formed.

Proteins will interact with DNA in each of these ways and, also combinations of the two are known. We shall restrict ourselves to describing only one of each type: the binding of DNA with the enzymes pancreatic deoxyribonuclease I and *Eco* RI and with the regulatory CAP protein.

SAQ 1.8

The diagram below show the orientation of the basepairs in DNA in relation to the major and minor grooves of the double helix. Which chemical groups of the bases would be potentially capable of hydrogen bonding with a protein molecule? Note for simplicity we have omitted the carbon atoms of these bases.

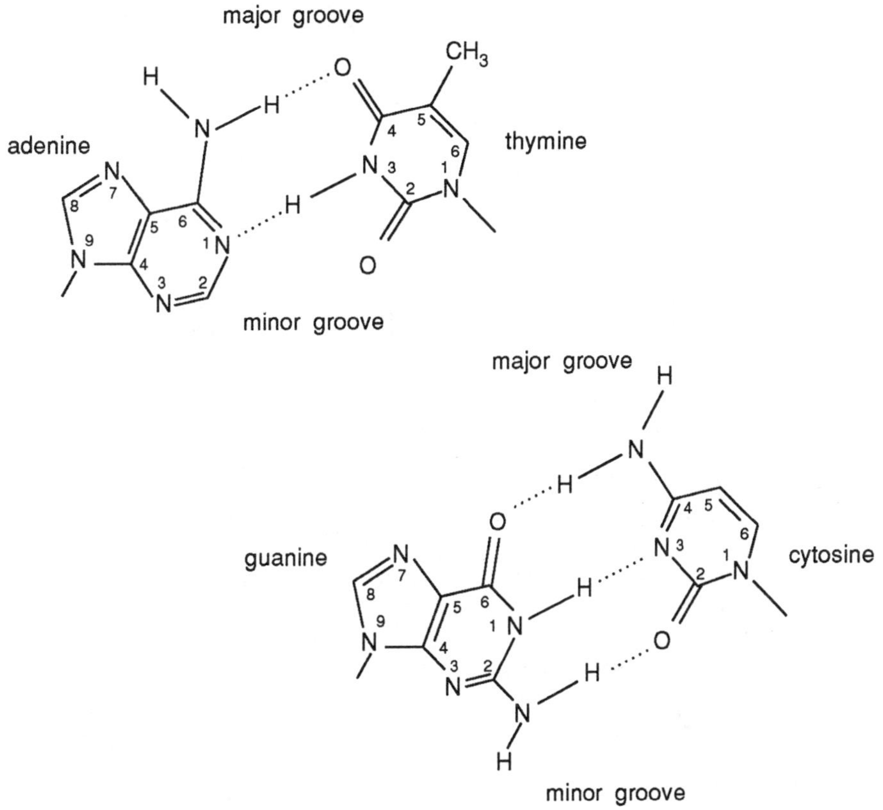

major groove

adenine   thymine

minor groove

major groove

guanine   cytosine

minor groove

Give your response by drawing a table of four columns to show the groove concerned, the base involved, the chemical groups associated with the base and whether they are donors or acceptors of hydrogens in hydrogen bonding.

1.4.1 Deoxyribonuclease I

DNase I   Deoxyribonuclease I (DNase I) is a hydrolytic enzyme which digests DNA into oligonucleotides with an average length of four nucleotides (Examine Figure 1.12). The structure of the protein has been investigated and its mode of interaction with DNA determined. As you may expect, the enzyme does not interact with the bases of DNA. DNase I contains a number of Arg and Lys residues which form ionic links with phosphate groups on either side of the minor groove over a length of about one turn of the double helix. Thus binding, and subsequent hydrolysis of DNA, occurs in a manner independent of the base sequence.

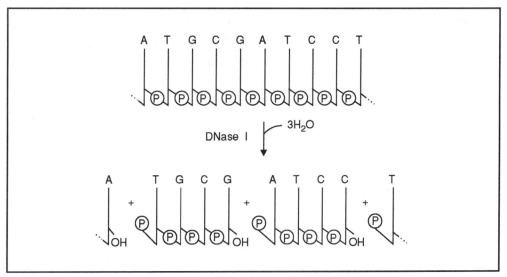

Figure 1.12 Hydrolysis of DNA by DNase I.

1.4.2 *Eco*RI

*Eco*RI   Restriction endonuclease enzymes hydrolyse DNA in a manner which is base-sequence dependent. Different restriction endonucleases recognise different short base sequences. All of the sequences recognised are palindromes, ie the sequences of bases in each of the strands is the same but runs in the opposite direction (Figure 1.13a). (Note a palindrome is a term that is used to describe a word that is spelt the same from either end, for example, radar).

Eco RI recognises the hexanucleotide sequence shown in Figure 1.13a in double-stranded DNA and cleaves both strands at the positions indicated by the arrows.

Like its substrate, *Eco* RI is a symmetrical molecule, and is composed of two identical subunits. Binding of the enzyme distorts the structure of DNA (Figure 1.13b), widening the major groove and allowing the enzyme molecule greater access to, and therefore improved recognition of, the bases within the double helix.

Basepairing inside the distorted region is not disrupted, even though the specific hexanucleotide substrate region is recognised by the formation of 12 precise hydrogen bonds between the enzyme molecule and the DNA bases. Residue Arg-200, of each subunit, forms two hydrogen bonds with a guanine, while residues Glu-144 and Arg-145 each form two hydrogen bonds with the adjacent adenine bases (Figure 1.13c).

It is unlikely that *Eco* RI could bind directly to an appropriate hexanucleotide sequence. A more plausible suggestion is that it initially binds in a fairly random manner to the sugar-phosphate backbone by electrostatic interactions. The enzyme then migrates along the major groove, causing the local distortion in the region of binding, until the appropriate sequence is located.

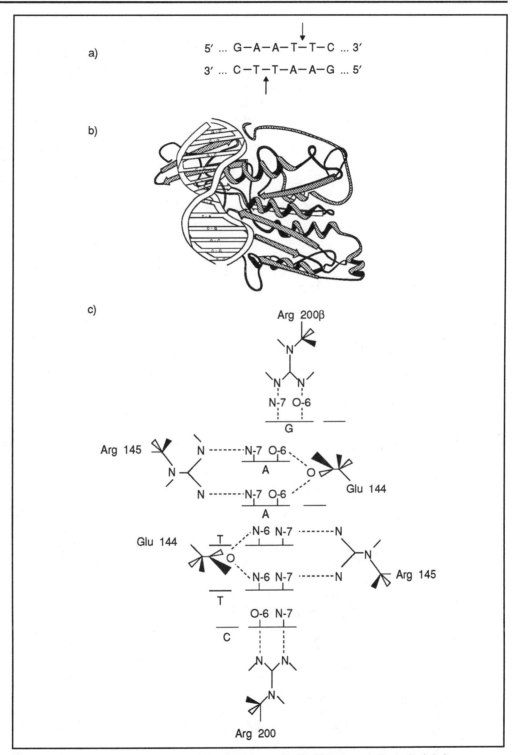

Figure 1.13 a) Palindrome site recognised by the enzyme *Eco*RI. The arrows indicate the bonds hydrolysed. b) Molecular model showing the binding of one subunit of *Eco*RI to double helical DNA. Note the distortion induced in the DNA. c) Hydrogen bonds which stabilise the DNA-*Eco*RI complex are shown as dotted lines. Figures b) and c) redrawn from Rosenberg J M *et al* (1987) TIBS 12, 395-398.

1.4.3 The CAP protein

cap protein

The regulation of metabolism means that a number of regulatory proteins must bind to DNA in order to switch off or on specific genes. A number of such regulatory proteins are known. One, the CAP protein of *E. coli*, is involved in regulating the production of enzymes involved in glucose metabolism. It has a structure typical of a number of DNA-binding regulatory proteins in having a so-called helix-turn-helix motif as part of its molecular structure. The side chains of the amino acid residues in the 'second helix' can associate with target bases in the DNA double helix, which in turn leads to regulation in the expression of the relevant gene. The association between the CAP protein and DNA is stabilised by an elaborate system of hydrogen bonding, van der Waals interactions and salt links which ensures precise and specific binding.

1.4.4 Other proteins which bind to DNA

We will be considering the synthesis of RNA using DNA as a template in later chapters. This process, called transcription, requires the binding of transcription enzymes to specific sites along the DNA. You should also be aware that a wide variety of proteins bind to DNA in eukaryotes. These are discussed in the BIOTOL text 'Genome Management in Eukaryotes'.

1.5 Bacterial chromosomes

bacterial chromosomes

Bacteria and viruses contain a single, or relatively small number of, chromosomal nucleic acid molecule(s). In the majority of species these are double-stranded DNA. However, some viruses have single-stranded DNA or RNA, which may be circular or linear depending upon the viral type (see Table 1.3).

Examination of *E. coli* cells by electron microscopy shows that is chromosome occupies an irregularly-shaped space, relatively free of ribosomes, called the nucleoid (Figure 1.14a). The nucleoid contains a double-stranded molecule of DNA of diameter about 350 µm, radius 2-4 nm and mol wt 2.5×10^9. This is equivalent to a molecule 1.1 mm long. It is not, however, fully extended in the bacterial cell - nor could it be since *E. coli* cells are only 1-2 µm long!

If the nucleoid of *E. coli* is extracted from the cells using mild procedures into solutions rich in cations, then it remains in a highly condensed state. The chromosome consists of 50-100 loops, each loop formed of a supercoiled section which greatly reduces the overall length of the DNA molecule, allowing it to be packaged within the limited confines of the cell (Figure 1.14b).

basic proteins

The loops are stabilised by the presence of small basic proteins. One protein, the DNA-binding protein II, has a mol wt of 9500 and binds to the DNA as a dimer. It has extended Arg-rich 'arms' which are thought to interact with the phosphate groups of the DNA backbone. One turn of supercoiled DNA would appear to consist of 80-100 nucleotide bases and bind 8-10 dimers of binding protein II.

| SAQ 1.9 | What is the ratio of the length of the *E. coli* chromosome to the length of its cells? |

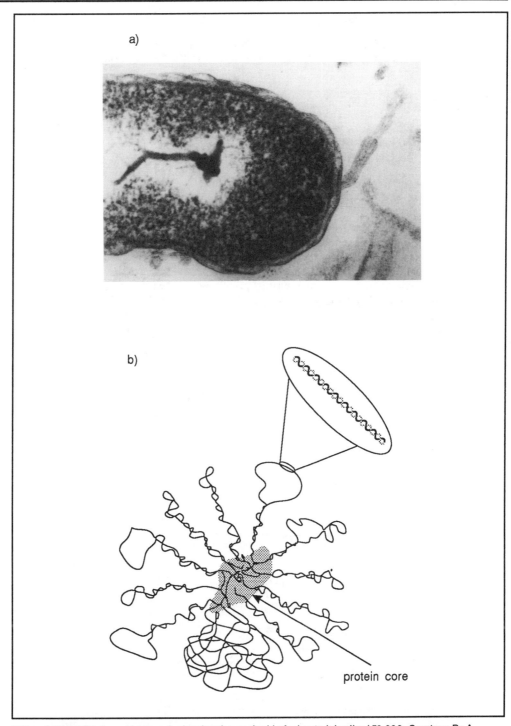

a)

b)

protein core

Figure 1.14 a) Electron micrograph showing the nucleoid of a bacterial cell x 150 000. Courtesy Dr A Curry, Public Health Laboratory, Withington Hospital, Manchester. b) Schematic representation of an isolated bacterial chromosome. For the sake of clarity only a few of the 50-100 loops have been drawn and the DNA-binding proteins which help stabilise the loops have been omitted. Redrawn from Kellenberger E (1990) 'The Bacterial Chromosome' (eds Drlica K and Riley M) p180. Published by the American Society for Microbiology, Washington, DC.

1.5.1 Plasmids

Plasmids are extra chromosomal DNA molecules mainly found in bacteria. They are relatively small, double-stranded circular DNA molecules that code for only a few genes.

F, R and col
factors

E. coli cells may contain three types of plasmids which are the F (for fertility), R (resistance) and col (colicinergic) factors. F factors promote sexual conjugation in *E. coli*, while the R factors confer resistance to some drugs. The col factors give the bacteria possessing them the capacity to secrete colicins. These are antibiotics which kill bacterial cells lacking the col plasmid.

1.6 Replication of chromosomal DNA

The structure of DNA proposed by Watson and Crick immediately suggested how genetic information (ie the DNA) could be replicated and so passed to the subsequent generation. If the two strands were separated, each could form a template allowing new daughter double-stranded molecules to be synthesised following the usual basepairing rules. Such a process would generate two daughter DNA molecules, each identical to the original (ie parental) DNA. Thus each of the DNA molecules produced would consist of one parental strand and one daughter strand. This type of DNA replication is called semi-conservative (Figure 1.15).

semi-
conservative
replication

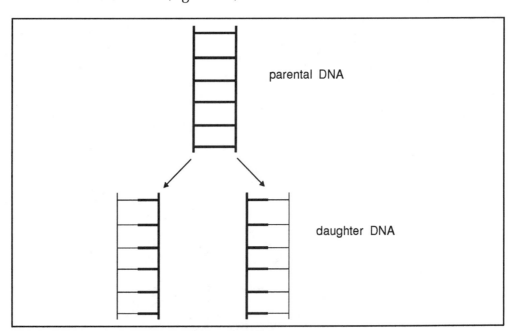

Figure 1.15 Schematic diagram of semi-conservative replication of DNA.

Semi-conservative replication of DNA involves the addition of new deoxyribonucleotides to the 3' end of the growing strand of the DNA molecule (Figure 1.16). The addition is catalysed by the enzyme DNA polymerase (DNA pol) and requires the four deoxyribonucleotides (dATP, dGTP, dCTP and dTTP) Mg^{2+}, a template and a primer.

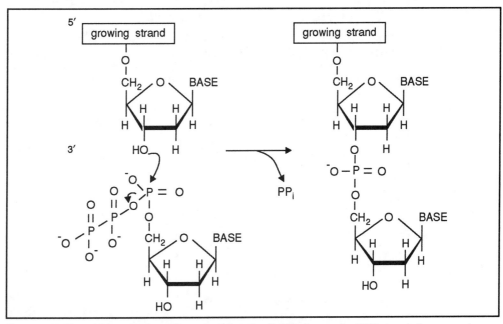

Figure 1.16 The addition of a deoxyribonucleotide to the 3'-OH of a growing DNA strand. A phosphodiester bond is formed and pyrophosphate (PPi) is released.

The template is the strand of DNA which is to be copied and which aligns the new nucleotides in the correct order according to the normal complementarity rules. The role of the primer will be discussed below. The equation:

$$(DNA)_n + dNTP \rightleftarrows (DNA)_{n+1} + PP_i$$

summarises the addition of a deoxyribonucleotide to the growing strand. PPi is pyrophosphate. Its subsequent hydrolysis makes the addition energetically favourable.

1.6.1 Replication of bacterial DNA

An elegant experiment by Meselsohn and Stahl confirmed that bacteria replicate by a semi-conservative mechanism. They grew *E. coli* in a medium containing a heavy isotope of nitrogen ($^{15}NH_4C1$). Bacteria can synthesise all their purine and pyrimidine bases using this source of nitrogen. The bacteria were allowed to grow for several generations so that all their DNA was labelled with ^{15}N giving so called 'heavy' DNA. DNA containing only ^{14}N is called 'light' DNA. The two types of DNA may be separated by centrifugation. The bacteria were then transferred to a medium containing $^{14}NH_4Cl$ and allowed to divide once. The DNA was isolated and subjected to centrifugation in a gradient of CsCl, a technique which separates molecules according to their density. The DNA isolated was found to have a density corresponding to an atomic weight for nitrogen of $^{14.5}N$. This indicated that one strand of the DNA was derived from the parental DNA (ie contains ^{15}N) and the other strand was newly synthesised (ie has ^{14}N).

SAQ 1.10

If *E. coli* were allowed to reproduce twice in the presence of $^{14}NH_4Cl$, following growth in $^{15}NH_4Cl$, what would be the relative densities of the DNA molecules produced?

1.6.2 Overview of replication in *E. coli*

origin of
replication

Replication of the *E. coli* chromosome always begins at the same position, the origin of replication or *oriC*. This is a sequence of 245 basepairs which is recognised by proteins that initiate replication. Replication beings simultaneously in both directions, and occurs at the same rate around the circular chromosome. This produces a so called θ-structure (Figure 1.17a).

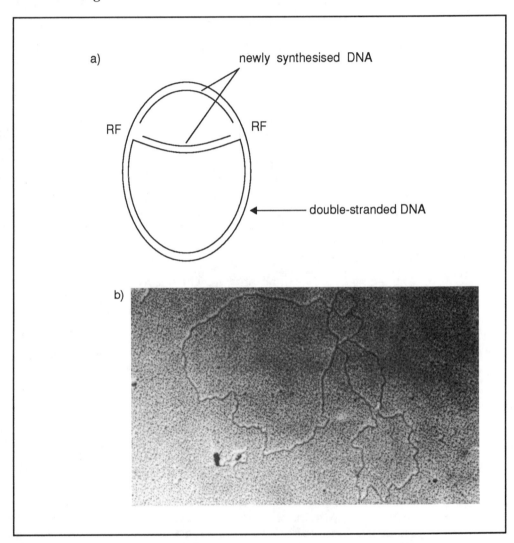

Figure 1.17 a) The replication of a circular double-stranded DNA molecule generates a so called θ structure (named after its resemblence to the Greek letter θ). RF = replication fork. b) Electron micrograph of a replicating circular DNA molecule. The two replication forks are indicated by arrows. Courtesy Dr P A Kumar, Department of Biological Sciences, Metropolitan University of Manchester.

The site of DNA synthesis is called a replication fork and because replication proceeds in both directions there are two of these on a replicating chromosome. Replication ends when the two replication forks meet at a termination site, opposite *oriC*, called *terC*. Figure 1.18 shows the events occurring at a replication fork and the following describes this in more detail.

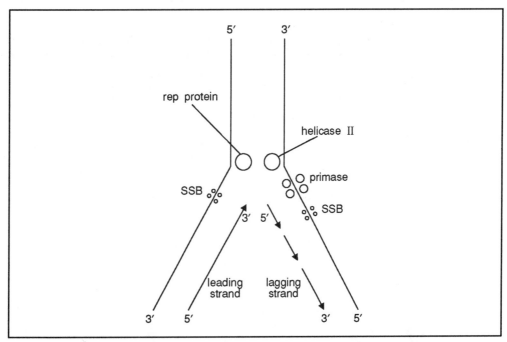

Figure 1.18 A replication fork showing some of the proteins invovled in replicating *E. coli* DNA. Rep is a helicase; SSB, single-strand binding proteins. See text and Table 1.5 for details.

1.6.3 Unwinding of the double helix

helicases

In order to separate the two strands of the parental DNA at the replication fork, enzymes called helicases are required. Helicases bind to the template strands of DNA and use the free energy of hydrolysis of ATP to move along the strands separating the double helix. One helicase, the rep protein attaches to the $5' \rightarrow 3'$ template strand and moves in the $3' \rightarrow 5'$ direction. The other helicase (sometimes called helicase II) moves in the $5' \rightarrow 3'$ direction along the $3' \rightarrow 5'$ parental DNA strand (Figure 1.19). Once unwound, single-strand binding (SSB) proteins complex with the single-stranded DNA and stabilise it in a rigid, extended conformation ideal for copying.

single-strand binding proteins

1.6.4 The role of primer RNA

RNA primer

DNA pol enzymes cannot start synthesis *de novo* and can only add nucleotides to a free 3'-OH. The primer used in DNA synthesis is a short piece of RNA containing 2-5 nucleotides. RNA synthesis can be started without a primer, and the RNA formed will have a free 3'-OH to accept the first deoxyribonucleotide. The RNA is synthesised by an enzyme called a primase which combines with several other proteins to form a complex structure called a primosome. Why is RNA rather than DNA used as a primer? One reason may be that DNA must be copied with great accuracy. Only about one base in 10^8-10^{10} is miscopied during DNA biosynthesis. This is an error rate sufficient to allow variation without endangering the viability of the organism. The greatest possibility of errors occurs when the first few bases are being laid down at the start of synthesis. The error rate can be minimised by initially incorporating ribonucleotides (eg the primer) and then removing the RNA at a later stage. This makes a high initial degree of accuracy unnecessary since mistakes can be corrected when the RNA is replaced with deoxyribonucleotides.

1.6.5 The role of DNA polymerases

DNA
polymerases

Three different DNA pol enzymes have been isolated from *E. coli* and are numbered I, II and III. DNA polymerase III is the enzyme responsible for the elongation of DNA. It consists of seven subunits called α, β, γ, δ, ε, τ, and θ. The enzyme occurs as a dimer at the replication fork. Pol III adds new deoxyribonucleotides to the free 3′-OH of the primer and therefore synthesises new DNA in the 5′ → 3′, direction. The α subunit is responsible for the synthetic activity.

Pol III can synthesise DNA only in the 5′ → 3′ direction. One parental strand can thus be copied continuously (the 3′ → 5′ strand). This copying produces the so called leading strand (Figure 1.19).

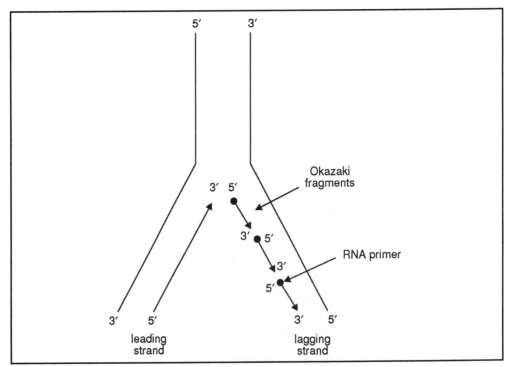

Figure 1.19 The synthesis of the 5′ → 3′ strand of DNA generates a continuous leading strand while that of the 3′ → 5′ produces the Okazaki fragments of the discontinous strand.

Okazaki
fragments

The other parental strand (the 5′ → 3′) must be partly unwound before copying can begin to produce the so called lagging strand. The lagging strand is synthesised as numerous small sections of DNA called Okazaki fragments, each about 500-2000 deoxyribonucleotides long. The formation of each Okazaki fragment requires an RNA primer (Figure 1.19). Thus the primase will have to be in continuous contact with the lagging strand. In contrast, the production of the leading strand requires only a single primer. Synthesis of an Okazaki fragment ceases when it comes in contact with the 5′ terminus of the preceeding fragment.

SAQ 1.11	DNA pol III can add deoxyribonucleotides to a growing DNA chain at about $800\ s^{-1}$. What is the minimum time required for the complete replication of the *E. coli* chromosomal DNA molecule? Remember that the genome of this organism is about 1.1 mm long and that a basepair occupies about 0.34 nm.

DNA pol I has a general role in the maturation of the newly formed DNA strands. It has $5' \rightarrow 3'$ exonuclease activity which removes the primer RNA. Pol I then fills the gap created by removal of the RNA with deoxyribonucleotides complementary to the parental DNA. The separate pieces of each newly completed DNA strand are then joined by the enzyme DNA ligase.

The biological function of DNA pol II is unknown.

1.6.6 Proofreading during DNA synthesis

DNA polymerases initially incorporate incorrect bases at the relatively high frequency of one base in 10^5, but newly synthesised DNA differs from parental DNA in having an error rate of only about one base in 10^8-10^{10}. There must therefore be a mechanism for checking the accuracy with which the DNA is copied and remedying any mistakes. This

proofreading   is called proofreading. The ε subunit of pol III is a $3' \rightarrow 5'$ exonuclease ie it degrades DNA by removing nucleotides from the 3' end of DNA. Pol III has a proofreading capacity linked to its $3' \rightarrow 5'$ exonuclease activity.

DNA pol III   DNA polymerase III contains a deep cleft which binds DNA. The DNA may move either forward or backward relative to the enzyme. If the DNA contains a mismatched base the DNA will be distorted preventing its forward movement. Under these circumstances the DNA will diffuse backwards coming into contact with the active site associated with $3' \rightarrow 5'$ exonuclease activity and the offending base is removed. DNA pol III then inserts the appropriate base and forward movement re-commences.

∏ What would be the biological consequences if DNA was copied with complete accuracy?

It would reduce variation in those organisms in which mutations are the sole way that genetic variability can occur.

1.6.7 The role of topoisomerases

Unwinding of a circular double-stranded DNA molecule will dissipate its normal negative supercoiling and eventually generate positive supercoils in the DNA (see earlier). Eventually the strain produced will halt movement of the replication fork and so prevent replication. However, this impediment is removed by topoisomerases which

topoisomerases   can catalyse changes in the amount of supercoiling. Topoisomerase I makes single-stranded breaks in DNA, so the unreplicated DNA can rotate relative to the replication fork releasing the torsional stress and allowing replication to proceed.

concatenated DNA   Complete replication of a double-stranded circular DNA molecule results in the formation of two interlinked ('concatenated') DNA daughter molecules. These are separated by Topoisomerase II which makes breaks in both strands of the DNA, unlinks the molecules and then reseals the breaks. This allows the two DNA molecules to separate and to be segregated between the two new cells.

Π   The replication of bacterial chromosomal DNA is dependent upon the activity of a number of proteins which bind to DNA in precise and specific ways. Table 1.5 lists these proteins, together with their mol wt and the number of molecules found on average in the cells of *E. coli*. The table is, however, incomplete! You may find it helps your understanding of replication to complete it, using the material given in Sections 1.6.2-1.6.7.

Protein	mol wt x 10^{-3}	molecules per cell	function
primase	60	50	
rep (helicase I)	65	50	
helicase II	300	20	
SSB	74	500	
DNA pol III	800	200	
topoisomerase I	100		
DNA gyrase	400	250	
DNA pol I	102	300	
DNA ligase	74	300	

Table 1.5 Some of the proteins involved with DNA replication in *E. coli*.

The functions you should have listed in Table 1.5 are successively: synthesises primer; unwinds double-stranded DNA; unwinds double-stranded DNA; stabilises single-stranded DNA; synthesises DNA; relaxes supercoiled DNA; induces supercoiling; maturation of DNA; joins free ends of DNA;

1.6.8 The bacterial cell cycle

The bacterial cell cycle is defined as the period between successive bacterial divisions. For *E. coli* it is about 20 minutes. Bacteria replicate their DNA rapidly: although it takes about 40 minutes to completely replicate the *E. coli* chromosome. How can we explain the ability of the bacterium to divide in a shorter time that it takes to synthesise its chromosomal DNA? More than one round of DNA synthesis is occurring at any one time. This can be explained because a second round of DNA synthesis may begin before the first replication fork reaches its terminus. The cell cycle is therefore shorter than the time required to replicate DNA!

SAQ 1.12   How many chromosomal DNA molecules per cell would you expect to find in a rapidly growing culture of *E. coli*.

1.7 Replication of viral nucleic acids

viral replication   Viruses have a great diversity of nucleic acids (Table 1.3). Also, all viruses are obligate parasites and can only replicate in the cell they infect. Thus they show a variety of mechanisms for replicating their nucleic acid.

1.7.1 Circular double-stranded DNA molecules

DNA viruses Circular viral DNA molecules may replicate by one of two methods. They may use a mechanism similar to that found in *E. coli* forming a θ-intermediate. Alternatively, they may use a rolling circle mechanism.

One strand of the DNA (called the + strand) is nicked by DNA polymerase generating a free 3'-OH. Nucleotides are added to the 3' end and, at the same time, the 5' end is displaced or rolled off the molecule (Figure 1.20). SSB proteins are involved in stabilising the extended strand. The complementary, - strand is copied as Okazaki fragments. A 'rolling circle' mechanism presents no topological problems and topoisomerases are not involved in this type of replication.

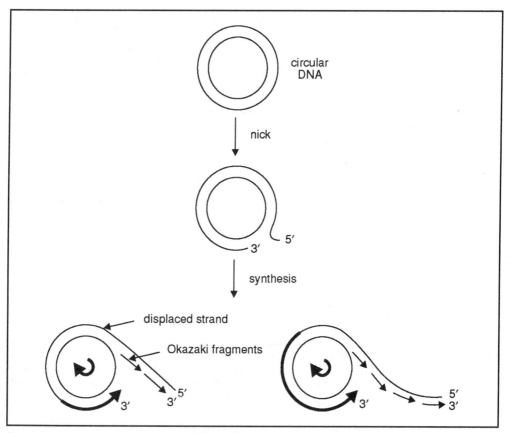

Figure 1.20 The rolling circle model of viral DNA replication.

Synthesis can stop after one replication or continue around the circle several times producing concatemers. The concatemers consist of multiple copies of the genome linked end to end. These can be cut into individual genomes and then their free ends joined together by DNA ligase to produce circular molecules.

1.7.2 Linear double-stranded DNA molecules

Some linear double-stranded DNA molecules can be replicated without the need of a primer. Figure 1.21 shows an example of such a process in the virus, bacteriophage φ29.

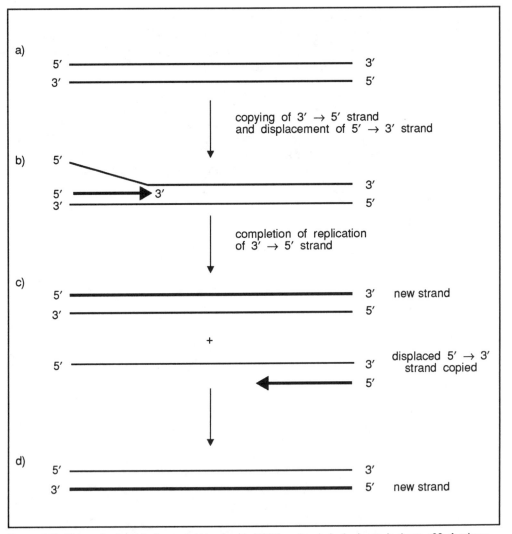

Figure 1.21 The replication of a linear double-stranded DNA molecule in the bacteriophage φ29. A primer RNA molecule is not required. a) Linear double-stranded DNA, b) one strand is copied while the other is displaced, c) the first strand is completed, whilst copying of the second strand commences, d) strand synthesis is completed.

1.7.3 Linear single-stranded DNA

Linear single-stranded viral DNA can replicate in one of a number of ways. For example, the linear molecule can be converted into a circular form because it has sequences of bases at each end which are complementary to each other. On replication concatemers are formed which are later cleaved to release single-stranded daughter DNA molecules.

1.7.4 Circular single-stranded DNA

The phage φX-174 contains a circular single-stranded DNA molecule (+ strand). This is converted into a double-stranded form using an RNA primer and pol III. The removal of the primer and filling of the gap is carried out by pol I. New + strands are then generated by a rolling circle method as described above.

1.7.5 Replication of viral RNA

Viruses which use RNA as their genetic material must use either an RNA-dependent RNA polymerase or an RNA-dependent DNA polymerase.

RNA-dependent RNA polymerase viruses

Class 1 viruses contain + RNA genome molecule. A replicase makes a complementary copy called - RNA which is used as a template to make new + RNA for new viral particles.

Class 2 viruses have a - RNA. The virus particle also has an RNA replicase which makes a + RNA copy. The + RNA can serve as a template for producing new - RNA molecules.

Class 3 viruses contain multiple copies of double-stranded ∓ RNA. On infecting the cell the virus transfers an RNA polymerase along with the RNA. The polymerase copies only the +strand. This associates with specific proteins and then serves as a template for the production of -strands for the viral progeny.

RNA-dependent DNA polymerase viruses

reverse transcriptase

Class 4 viruses have a RNA-dependent DNA polymerase (a reverse transcriptase). They contain single-stranded RNA which is replicated via a DNA intermediate. Viruses of this type are called retroviruses and are responsible for causing some forms of cancer in animals and AIDS in primates.

retroviruses

Retroviruses make a DNA copy of the RNA using reverse transcriptase. The RNA is then degraded and the DNA is copied to make a double-stranded molecule (Figure 1.22). This becomes incorporated into the host DNA and is used to make copies of viral RNA for new virus particles.

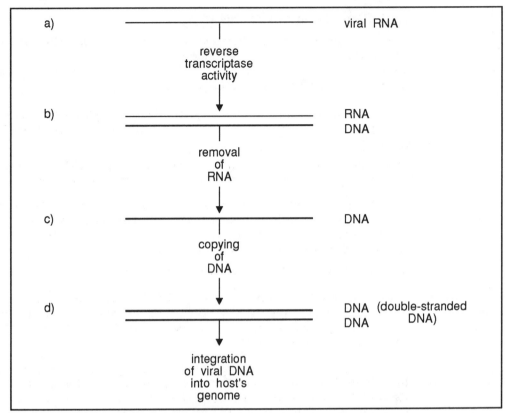

Figure 1.22 The replication of viral RNA. The reverse transcriptase makes a DNA copy complementary to the RNA strand a). The RNA is then degraded c) and the DNA copied to make a double-stranded molecule d). This is then integrated into the host's genome and used to produce viral RNA.

SAQ 1.13

Indicate the following as true or as false. Briefly explain your response.

1) Replication of viruses can only occur in cellular organisms.

2) RNA viral replication always involves RNA-dependent RNA polymerase activity.

3) Replication of viral RNA always involves the nucleotides ATP, GTP TTP and CTP.

4) Viruses containing single-stranded RNA always make a complementary DNA copy of it during their replication cycle.

5) RNA viruses always contain a + RNA molecule.

Summary and objectives

Cellular organisms use DNA as their genetic material, although viruses use DNA or RNA for this purpose. DNA is composed of deoxyribose, the bases adenine, guanine, cytosine and thymine and phosphate. RNA contains ribose, not deoxyribose, and uracil rather than thymine.

In DNA the deoxyribonucleotides are arranged in two antiparallel chains held together by hydrogen bonds. DNA can assume one of a variety of double helical structures of which B-DNA is the most widespread.

The sequence of bases in one strand of the double helix constitutes the genetic information. This must be copied with great accuracy during replication and, indeed, forms the template to direct the formation of daughter DNA molecules. Thus replication is by a semi-conservative mechanism. Replication of bacterial chromosomal DNA requires the co-ordinated activity of a large number of proteins. The roles of many of these proteins are well understood.

Viruses can only replicate in a host cell. Therefore the mode of replication depends upon the host cell and the type of viral nucleic acid involved.

Now that you have completed this chapter you should be able to:

- list the components of DNA;

- describe/explain the double helical structure of DNA, with emphasis on B-DNA;

- describe the range of structures exhibited by DNA;

- explain the need for careful packaging of DNA within cells;

- describe the packaging of chromosomal DNA in prokaryotic cells;

- outline in general terms the ways in which a variety of protein molecules are able to interact with DNA;

- describe the way in which the prokaryotic chromosome is accurately replicated and identify the roles of the proteins involved;

- outline the replication of viral nucleic acids including that of retroviruses.

Mutagenesis

Mutagenesis

2.1 Introduction

nutrition

A mutation is a heritable change in the information stored in the genome. The sequence of bases in the DNA of an organism constitutes the genetic information. The sequence of bases form the genetic code that can be translated in a messenger RNA into a sequence of amino acid residues in a polypeptide. Alterations in the genetic material

evolution and adaption

regularly occur and are the basis of evolution and adaptation of all organisms including micro-organisms. Genetic variation can be brought about in a number of ways the most important of which are mutation, recombination and transposition. This chapter will discuss only the first of these.

	U		C		A		G		
	UUU	Phe	UCU	Ser	UAU	Tyr	UGU	Cys	U
	UUC	Phe	UCC	Ser	UAC	Tyr	UGC	Cys	C
U									
	UUA	Leu	UCA	Ser	UAA	End	UGA	End	A
	UUG	Leu	UCG	Ser	UAG	End	UGG	Trp	G
	CUU	Leu	CCU	Pro	CAU	His	CGU	Arg	U
	CUC	Leu	CCC	Pro	CAC	His	CGC	Arg	C
C									
	CUA	Leu	CGA	Pro	CAA	Gln	CGA	Arg	A
	CUG	Leu	CCG	Pro	CAG	Gln	CGG	Arg	G
	AUU	Ile	ACU	Thr	AAU	Asn	AGU	Ser	U
	AUC	Ile	ACC	Thr	AAG	Asn	AGC	Ser	C
A									
	AUA	Ile	ACA	Thr	AAA	Lys	AGA	Arg	A
	AUG	Met	ACG	Thr	AAG	Lys	AGG	Arg	G
	GUU	Val	GCU	Ala	GAU	Asp	GGU	Gly	U
	GUC	Val	GCC	Ala	GAC	Asp	GGC	Gly	C
G									
	GUA	Val	GCA	Ala	GAA	Glu	GGA	Gly	A
	GUG	Val	GCG	Ala	GAG	Glu	GGG	Gly	G

Table 2.1 The genetic code. The first nucleotide for each triplet is given in the left hand column and the second nucleotide is given across the top and the third on the right. Note that three codons are terminal codons that specify the end point of translation.

A protein can contain up to 20 different amino acids. Each amino acid is coded for by a codon of three bases. This genetic code is shown in Table 2.1. The code is described in terms of the messenger RNA which transfers information between DNA and protein. In RNA, uracil is used rather than the thymine of DNA.

Mutation may produce only a slightly altered variant of the protein and biological function may be retained. It may be difficult to detect such alterations other than by biochemical analysis of proteins. Other mutations may lead to noticeable changes in appearance, physiological function or even to the death of an organism. Thus mutations may produce both phenotypic and genotypic changes.

A nomenclature for distinguishing between phenotype and genotype has been established. If for example a bacterium lacks an enzyme arginase the genotype is described as arg^-, the superscript indicating that the enzyme deficiency is due to a mutation. The phenotype is designated Arg^-. Table 2.2 lists some commonly mutated genes of *Escherichia coli*.

Gene	Function	Gene product
lacZ	LACtose metabolism	β-galactosidase
trp	TRYPtophan synthesis	Trypthophan synthetase
rp1A	ribosomal protein	ribosomal protein L
polA	DNA POLymerase	DNA polymerase 1
leuA	LEUcine biosynthesis	β-isopropylmate synthesis
pyrG	PYRimidine synthesis	CTP synthesis

Table 2.2 Genes commonly mutated in *E. coli*. We have used upper case lettering in part of the function. These letters give a clue as to the phenotype description of organisms with or without the gene.

SAQ 2.1

A bacterium has a mutation in the gene coding for a β-galactosidase such that the enzyme is no longer functional. Use an accepted nomenclature to describe the genotype and phenotype.

Before we discuss the various types of mutation, we need to have a clear idea about the polarity of DNA, RNA and protein synthesis and some of the conventions used in describing the two strands of DNA.

Consider double-stranded DNA:

$$5'\ \ TTCA \ \text{———} \ 3'$$
$$3'\ \ AAGT \ \text{———} \ 5'$$

One of the strands carries the genetic information and is transcribed by the DNA-dependent RNA polymerase to produce messenger (m)RNA.

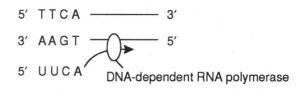

5′ T T C A ——————— 3′

3′ A A G T ——○——▶ 5′

5′ U U C A   DNA-dependent RNA polymerase

Note that the RNA is synthesised from the 5′ end and has a base composition complementary to the 3′ → 5′ strand of DNA. Note also that the nucleotide sequence of the mRNA is the same as that of the 5′ → 3′ strand of DNA except that U replaces T. The strand of DNA that is transcribed is called the sense (or coding) strand while the other strand is called the nonsense (or non-coding) strand. The conversion of the nucleotide sequence in mRNA into a sequence of amino acids during the process of translation is carried out in such a way that mRNA is read from the 5′ end and the amino acids are added to the protein beginning at the N (amino) end and finishing at the C (carboxylic acid) end. We will be examining transcription and translation in much more detail in later chapters.

∏ The coding strand of a DNA molecule has the nucleotide sequence 3′ AAA ATA GGT 5′ (note this sequence has been written 3′ → 5′). Write down the sequence of amino acids which will be coded for by this nucleotide sequence (use the information in Table 2.1).

You should have concluded that the amino acid sequence would be phe, tyr, pro. The mRNA that is made against the coding strand has the nucleotide sequence 5′ UUU UAU CCA 3′. From Table 2.1 this sequence codes for phe, tyr, pro. You will notice from this exercise that if you are given the sequence of nucleotide in the strand of DNA which is transcribed (the coding strand), then you have to go through two stages in order to determine the sequence of amino acids it codes for. It is much more convenient to write the nucleotide sequence of a gene in terms of the sequence of the non-transcribed (non-coding) strand as this will have the same sequence as the mRNA except T replaces U. We can then translate this sequence directly into a sequence of amino acids. Because of this simplification, molecular geneticists usually describe the nucleotide sequences of genes in terms of non-coding strands. In examining nucleotide sequences of genes therefore it is important to realise if the author has written the sequence of the non-coding or of the coding strand.

2.2 Point mutations

mutagens   Point mutations may occur naturally as the result of the incorporation of an incorrect base during DNA replication or as a result of the interaction of DNA with chemicals, called mutagens, which alter the bases of DNA.

A number of types of single base changes are possible. These are illustrated in Figure 2.1 and Table 2.3 summarises the major types of mutation and their effect on the gene. Examine these carefully as they contain a lot of information.

original gene (non-coding strand)

```
5' TTT  CCA  CTA  AGT      CGA  TGC  3'
   phe  pro  leu  ser      arg  cys
```

mutation

a)
```
T A T  CCA  CTA  AGT      CGA  TGC                    (transversion)
tyr    pro  leu  ser      arg  cys
```

b)
```
TTT  C T A  CTA  AGT      CGA  TGC                    (transition)
phe  leu    leu  ser      arg  cys
```

c)
```
TTT  CCA  CTA  GTC        GAT  GC                     (deletion)
phe  pro  leu  val        asp
```

d)
```
TTT  CC G ACT  AAG        TCG  ATG  C                 (insertion)
phe  pro  thr  lys        ser  met
```

e)
```
TTT  CC G GGA ATC  ACT  AAG  TCG  ATG  C              (insertion)
phe  pro  gly    ile   thr  lys  ser  met
```

f)
```
TTT  CCA  TGA ATC      CGA  TCG                       (inversion)
phe  pro  stop
```

Figure 2.1 The base sequence of part of a gene and its relationship to an amino acid sequence of a polypeptide. a) Replacement of a T by an A is an example of a transversion. b) Replacement of C by T is an example of a transition. c) Loss of a base (thymine) is a deletion. d) Addition of a base is an insertion. Note that the nucleotide sequences are those of the non-transcribed strand of DNA. (See text for fuller details).

transition

transversion
A transition is the change from one pyrimidine to another (C to T; or T to C) or from one purine to another (A to G; or G to A). A transversion changes a pyrimidine to a purine (T or C to A or G respectively) or vice versa (A or G to T or C respectively).

Type of mutation	Result
transition	G ↔ A or vice versa C ↔ T or vice versa
transversion	A or G ↔ C or T or vice versa
silent	new triplet codes for the same amino acid as the unmutated codon
neutral	mutated triplet codes for a chemically similar amino acid eg polar or hydrohobic
mis-sense	codes for different amino acid than the wild type
nonsense	chain termination
addition } deletion }	alters reading frame

Table 2.3 Genetic mutations and their cause.

SAQ 2.2	A gene contains the bases 5′ CGA AGT GGC GAT 3′ as part of its sequence. (Note this is the non-transcribed strand).

1) What sequence of amino acids is specified by these codons?

2) If a mutation caused a) A to G transition and b) a G to A transition, what would be the new sequences of amino acids?

2.3 Insertions and deletions

frameshift mutations

The insertion or deletion of a base alters the reading frame of the base sequence in the gene. These mutations are collectively called frameshift mutations (Figure 2.1 c and d). Frameshift mutations may lead to major changes in the amino acid sequence of a protein as every amino acid residue after the mutation will be altered.

silent mutations

Since the genetic code is degenerate (which means that there is more than one codon for an amino acid) it is possible for a mutation to occur without altering the amino acid composition of the protein. Such silent mutations are only detectable by chemical analysis of the bases of the DNA. Table 2.4 shows the nine possible mutations arising from a single base change in the tyrosine codon.

UAU (Tyr)		
CAU (His)	UGU (Cys)	UAC (Tyr)
AAU (Asn)	UUU (Phe)	UAA (Stop)
GAU (Asp)	UCU (Ser)	UAG (Stop)

Table 2.4 The effect of single base substitutions on the amino acid codon fo tyrosine.

∏   Which one of these mutations would be silent?

You should have spotted that the change of UAU to UAC would be a silent mutation.

There are a total of 549 (61 x 9) possible substitutions in the codons coding for amino acids. Due to degeneracy about a quarter of these fail to produce amino acid substitution.

SAQ 2.3	1) Examine the codons for the amino acids proline and threonine in Table 2.1. In each case there are four codons for the amino acid. List the similarities in these codons.

2) Mutations may lead to the following types of amino acid change:

   a) hydrophobic amino acid 1 to hydrophobic anino acid 2;

   b) polar amino acid 1 to polar amino acid 2;

   c) hydrophobic amino acid 1 to polar amino acid 2 or *vice versa* .

Which of these would you expect to be most deleterious? Give reasons for your choice.

2.4 Effect of a mutation on the gene

degenerate
code

The effect of a mutation on the function of the gene is dependent on the location of the altered base. The genetic code is degenerate, most amino acids are coded for by more than one codon. Where an amino acid is coded for by multiple codons the first two bases of the triplet (see Table 2.1) are often the same, and the third is A, G, C or U. Thus a mutation in the third base will change the codon but not the amino acid. This is an example of a silent mutation and will not give rise to an altered phenotype. Silent mutations can only be detected by analysis of the base composition of DNA.

missense
mutation

If a mutation occurs in position one or two of the codon then an amino acid change in the protein will occur. This is a mis-sense mutation and will lead to an altered phenotype. Whether the altered protein will be functional depends on the type of amino acid change and its position.

SAQ 2.4

If the mutation rate in bacteria is one in 10^6 per cell division and a culture contains 10^{12} cells, how many mutants might be present after one round of cell division.

Several different types of mutants are detectable.

2.4.1 Lethal mutants

lethal mutant

If a mutation occurs in a gene coding for an essential protein eg one of the enzymes required for energy metabolism or a protein controlling cell division then the organism will not be viable and will die.

2.4.2 Conditional lethal mutants

conditional
lethal mutants

Conditional lethal mutants result in death only under specific conditions. These mutants may contain, for instance, a heat sensitive enzyme which is inactivated at temperatures of $35° - 45°$ C. At lower temperatures ($20°$ C) the bacterium is prototrophic (like the wild type). Temperature sensitivity is an example of a conditionally lethal mutation, inducing non-viability only under special conditions. The mutant grows under permissive conditions and is killed under non-permissive conditions.

2.4.3 Auxotrophic mutants

auxotrophic
mutants

In an auxotrophic mutant, the gene coding for an enzyme involved in the synthesis of an essential nutrient eg a vitamin, amino acid or purine or pyrimidine base is mutated. The non-mutated organism is called the wild-type or prototroph and will be able to synthesise its nutritional requirements from a basic nitrogen and carbon source. The wild-type will grow on a minimal medium (Table 2.5). An auxotrophic mutant will grow on a minimal medium supplemented with the nutrient it is unable to synthesise.

Nutrient	Concentration (gl^{-1})
$KH_2 PO_4$	7.0
$KH_2 PO_4$	3.0
Na_3 citrate . $3H_2O$	0.5
$MgSO_4$. $7H_2O$	0.1
$FeSO_4$	1.0
glucose	2.0

Table 2.5 A typical minimal medium for growing *E. coli.*

Auxotrophs are important tools for exploring biochemical pathways and also provide evidence that each gene codes for an enzyme.

SAQ 2.5

Fill in the gaps in the following question using the list of words at the end.

A single base change in DNA is called a [   ]. A [   ] is the change from one [   ] to another for example (C → T) or one [   ] to another (A → G). A [   ] replaces a purine by a pyrimidine or vice versa [   ] or [   ] alter the reading frame of the gene and are examples of [   ] mutations. Mutations can have a number of effects on the product of the gene. If the codon is altered but not the [   ] the mutation is a [   ] one. A [   ] mutation results in the incorporation of a [   ] amino acid in the protein. If the mutated gene codes for an [   ] protein it may be a [   ] mutant. [   ] mutants contain for instance a heat sensitive enzyme that is inactivated at [   ].

An [   ] has one of the genes coding for the synthesis of an [   ] nutrient mutated. The non-mutant form is called the [-   ] and will grow on a [   ] medium.

Word list:

minimal; frameshift; deletions; point mutation; silent; essential (two times); amino acid; insertions; transition; wild-type; pyrimidine; 35°C - 45°C; mis-sence; lethal; auxotroph; transversion; purine; conditional lethal, different.

Π Beadle and Tatum were awarded the Nobel prize for a series of classic experiments with the bread mould *Neurospora crassa*. 1) Irradiation with X-rays produced mutant strains incapable of growing on a minimal medium lacking a particular amino acid. The mutant failed to grow on a minimal medium lacking tryptophan. 2) Crossing the mutant with the wild type produced spores half of which could grow on the minimal medium and half only could grow on the minimal medium if it was supplemented with tryptophan. What is the explanation for these observations?

Your explanation of these observations should be that:

1)   a mutation has affected the gene coding for an enzyme in an essential pathway leading to the synthesis of tryptophan;

2)   the mutation is in a gene inherited in a Mendelian fashion.

2.4.4 Constitutive mutant

constitutive and semi-constitutive mutants

Organisms which produce an inducible enzyme or group of enzymes for example the lactose metabolising enzymes of *Escherichia coli* (the *lac* operon) may mutate so that the enzymes are produced constitutively. Normally the lactose metabolising enzyme is only expressed in the presence of lactose (and absence of glucose). In constitutive mutants the enzymes are produced even in the absence of the substrate. In the case of the *lac* operon the constitutive mutant (I^-) produces an inactive repressor that no longer binds to the operator. Mutations may also occur in the operator (Oc) so that it binds the repressor less strongly than in the wild type. These mutants are called semi-constitutive. We will examine these regulatory aspects of gene expression towards the end of this text.

2.5 Mutant selection

Although it is relatively easy to generate mutants (see Section 2.10), their isolation can be difficult. The isolation of a mutant is called selection. A number of methods are available to isolate mutants from a mixture of bacteria.

2.5.1 Positive selection

positive selection

Mutants resistant to bacteriophages or antibiotics can be isolated from non-resistant strains by a process of positive selection. Such mutants can be grown in media containing the antibiotic or bacteriophage. Sensitive strains are killed whilst the resistant cells continue to grow. These resistant strains can be grown on plates (Petri dishes) containing the selective agent. Colonies derived from the mutant can be picked from the plate.

Auxotrophs cannot grow on a minimal medium but require enrichment and cannot be isolated by positive selection.

2.5.2 The isolation of auxotrophs

replica plating

The isolation of a specific auxotroph might appear to be insuperable considering only one bacterial cell in 10^{11} might show the mutation. The technique of replica plating allows the rapid screening of auxotrophic isolates. A master plate is prepared with a large number of separated colonies. A sterile sheet of velvet is pressed down on the master plate and cells are transferred to other plates containing a variety of different media (Figure 2.2). If for example a mutant lacked an enzyme of tryptophan biosynthesis and this amino acid was omitted from the medium in the replica plate then the cell transferred from the master plate would fail to grow. The colony on the master plate from which this cell was derived can then be identified and grown in the appropriate medium.

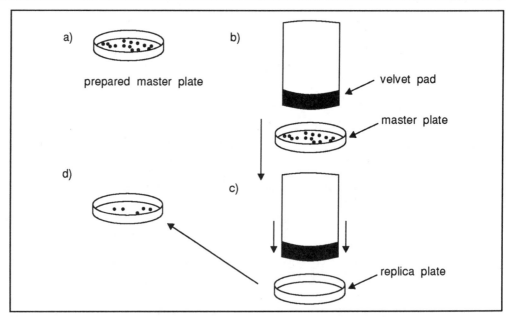

Figure 2.2 The technique of replica plating. a) A master plate is prepared with a large number of colonies some of which may be mutants, b) a sterile velvet pad is pressed onto the plate, c) and transferred to another plate with a minimal medium, d) the plate is incubated and the distribution of colonies compared with the master plate. Colonies which fail to grow on the replica plate are auxotrophic mutants and could be grown on a range of defined media.

2.5.3 Penicillin enrichment

penicillin
enrichment

Penicillin inhibits cell wall synthesis in growing bacteria but does not kill non-growing cells. For example if a bacterium was lac^- and placed on a medium containing lactose it would fail to grow whereas prototrophs able to utilise lactose would grow. Penicillin added to the medium would kill the growing prototrophs. The mutants can then be removed from the penicillin-containing medium and be grown in a medium containing a source of carbon other than lactose. Penicillin enrichment can be used if we can design growth conditions in which the wild-type cells grow, but the mutant cells cannot. After exposure to penicillin in these conditions, the proportion of mutants to wild-types will have increased because of the death of some of the wild-type cells. In other words, the viable cell population has become enriched with mutants.

2.5.4 Isolation of constitutive mutants

Constitutive mutants can be separated from the wild-type organism which only expresses the enzymes being studied in the presence of an inducer. In the case of the *lac* operon the inducer is a disaccharide. Several different methods have been used to isolate constitutive mutations in the *lac* operon and two examples will be given.

A substrate, 5-bromo-4-chloro-3-indoxyl-β-D-galactosidase (Xgal), can be degraded by the enzyme β-galactosidase releasing an indigo derivative which stains the cells blue. Constitutive mutants (I^-) express high levels of this enzyme and stain blue whereas the wild-type will not be induced by this substrate and the colonies will remain white.

Melibiose is an α-galactoside and *E. coli* requires the *lac* permease enzyme to take this sugar into the cell. Melibiose is a weak inducer of the gene coding for this enzyme in the wild-type and which do not produce high enough levels of the permease to grow on a melibiose medium. Constitutive mutation, however have high levels of the permase and growth is possible on melibiose medium.

A culture of *E. coli* K12 was subjected to the mutagenic effect of ultra violet light to induce auxotrophic mutants. The bacteria were grown in the presence of penicillin, 1) in a minimal salt medium. Mutants were identified by picking colonies and transferring to enriched agar plates, 2).

The specific growth factor requirements of two auxotrophs were determined by streaking onto 12 plates supplemented with various growth factors. The specific content of each of the plates is shown in the table below. Growth was found on plates 2) and 12) for mutant a) and on plates 3) and 5) for mutant b).

Plate number	Growth factor Supplements
1	adenine, biotin
2	biotin, riboflavin
3	phenylalanine, serine
4	serine, tryptophan
5	tryptophan, tyrosine, phenylalanine
6	tyrosine, nicotinic acid, threonine
7	threonine, alanine, riboflavin
8	alanine, arginine
9	arginine, proline, glutamic acid
10	proline, glutamic acid, aspartic acid
11	aspartic acid, biotin
12	riboflavin, nicotinic acid, biotin

Explain the purpose of steps 1 and 2 and identify the specific growth requirements of the two mutants.

2.6 The use of mutants for analysing metabolic pathways

Auxotrophs are particularly useful for analysing metabolic pathways since a large number of mutants are available and the mutation will lead to the loss of a specific enzyme.

The upper part of Figure 2.3 shows an example of a metabolic pathway which synthesises a compound D from A via intermediates B and C. Enzymes a, b and c carry out the specific reactions of synthesis. Mutants can be isolated deficient in enzymes a, b or c. If the mutant lacks enzyme a it will require B, C or D in the medium. A mutant lacking b will require C or D in the medium and a lack of c will require the organism to be fed with D.

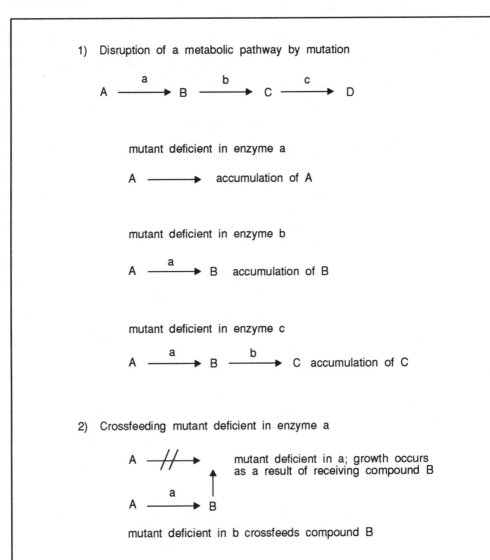

Figure 2.3 1) Auxotrophic mutants can be used to determine the sequence of metabolites in a metabolic pathway. 2) Crossfeeding, using a mutant to provide a substrate for a second mutant can confirm a proposed sequence (see text for discussion).

crossfeeding   A process of crossfeeding can help to order the steps of the pathway. A mutant deficient in a can be grown in the presence of a mutant deficient in b. The latter organism produces B required for growth by the first mutant. This is shown in the lower part of Figure 2.3.

Indicate which of the following statements are true and which are false?

1) Auxotrophs can be separated from prototrophs using an enriched medium containing penicillin.

2) Constitutive mutants continuously produce an inducible enzyme.

3) The process of isolating resistant mutants in the presence of the factor that the mutants are resistant to is called positive selection.

2.7 Methods of obtaining mutants

Bacterial mutants are routinely induced in the laboratory by treating bacterial cultures with chemicals called mutagens which cause base mispairings and by irradiation. However, as well as these induced mutations, some mutations occur spontaneously due to the structure of the bases in DNA. These spontaneous mutations will be discussed first.

2.7.1 Spontaneous mutations

spontaneous mutations

The frequency of spontaneous mutations in a gene is of the order of one in 10^5 - 10^7 cells per cell division. It is a common event and Table 2.6 shows the frequency of the event for some bacterial and fungal genes.

Organism	Phenotype	Gene	*Rate of mutation per cell division
E. coli	lactose fermentation	$lac^- \leftrightarrow lac^+$	2×10^{-7}
	lactose fermentation	$lac^+ \leftrightarrow lac^-$	2×10^{-6}
	histidine required	$his^+ \leftrightarrow his^-$	2×10^{-6}
	radiation resistant	$rad\text{-}s \leftrightarrow rad\text{-}r$	1×10^{-5}
Salmonella typhimurium	trp independent	$trp^- \leftrightarrow trp^+$	5×10^{-8}
Chlamydomonas reinhardii	streptomycin sensitive	$str^r \leftrightarrow str^s$	1×10^{-6}
Neurospora crassa	inositol required	$inos^- \leftrightarrow inos^+$	1×10^{-8}
	adenine independent	$ade^- \leftrightarrow ade^+$	4×10^{-8}

Table 2.6 The rate of spontaneous mutation in micro-organisms. *Note that the rate of mutation is expressed as number per cell division except for *Neurospora crassa*. Here the rate is expressed as number of mutations per asexually produced spore. Table redrawn from Lug *et al* 'Concepts of genetics' (1986). Merrill Publishing Co London.

Spontaneous mutations arise because the bases in DNA can assume a number of different structures. These different structures are called tautomers and are of two forms in DNA. Thymine and guanine normally adopt a keto structure (Figure 2.4) but can tautomerise to an enol form.

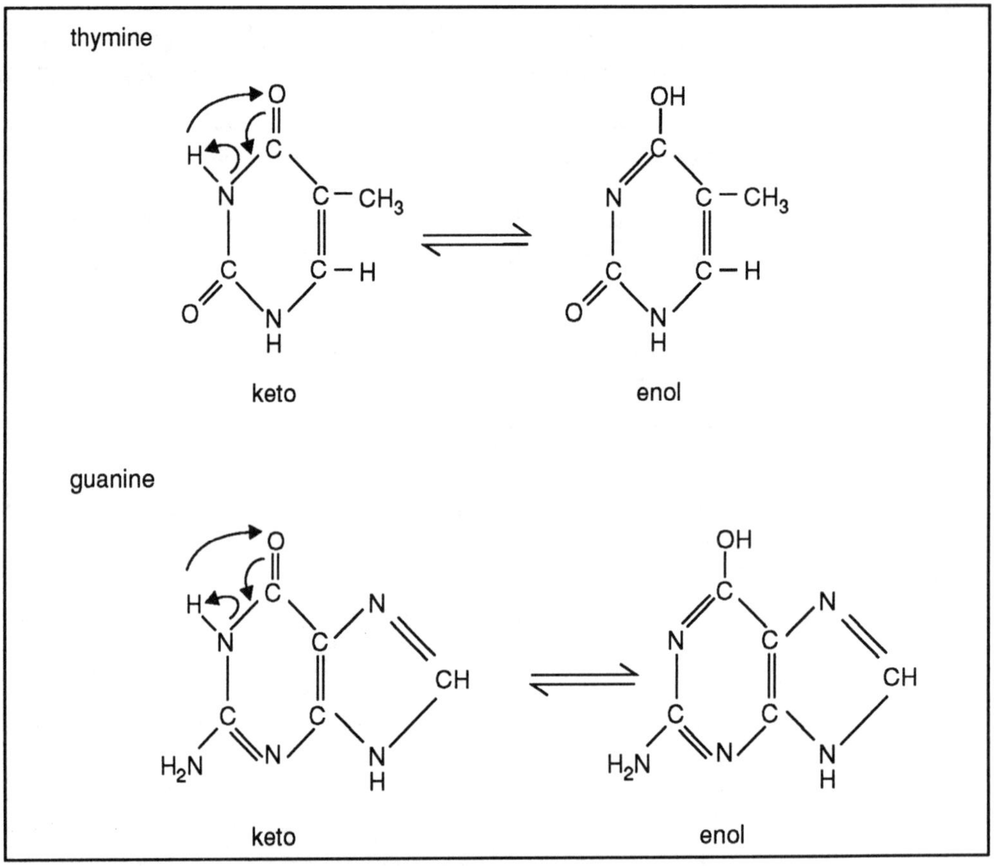

Figure 2.4 The structures of the keto and enol tautomers of thymine and guanine.

The enol tautomer allows thymine to pair with guanine (Figure 2.5).

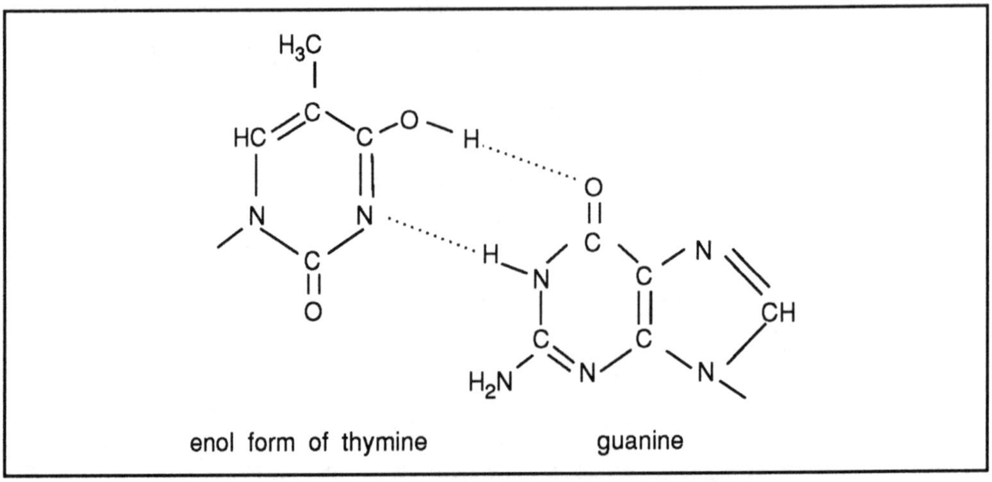

Figure 2.5 The enol tautomer of thymine can pair with the keto form of guanine.

The amino group of adenine and cytosine can tautomerise to an imino structure (Figure 2.6a) allowing an A-C pairing (Figure 2.6b) and the replacement of an A-T basepair by a G-C during DNA replication (Figure 2.7).

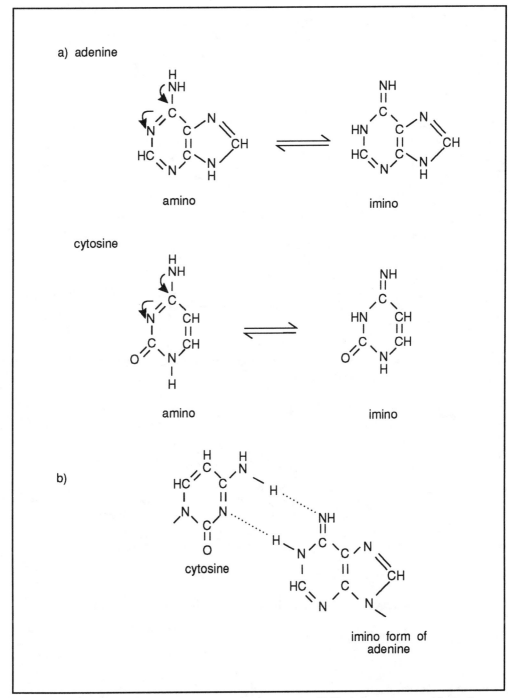

Figure 2.6 a) The amino and imino forms of adenine and cytosine. b) The imino tautomers of the purine base allow adenine to pair with cytosine.

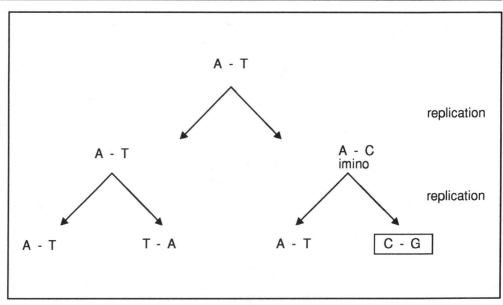

Figure 2.7 A transversion occurs when an adenine tautomerises to the imino form and pairs with cytosine.

Tautomerisation causes about one base in 10 000 to mispair and this would be expected to lead to a high rate of spontaneous mutation. However, such mutations are usually repaired during replication of DNA.

SAQ 2.8

Use Figure 2.7 as a model to demonstrate the consequences of the pairing of the enol tautomers of thymine and guanine. Begin with the basepair A-T.

2.7.2 Chemical-induced mutations

mutagens

Chemicals which cause mutations are called mutagens. Many mutagens are also carcinogens, that is they can cause cancer in eukaryotes. For this reason much study has been devoted to assessing the mutagenicity of chemicals.

Mutagens may be classified into one of three groups:

• analogues of bases of DNA;

• compounds which chemically modify bases;

• intercalating agents.

2.7.3 Analogues of bases found in DNA

base analogues   Base analogues for example 5-bromouracil, 5-iodouracil and 5-chlorouracil (Figure 2.8) are structurally similar to naturally occurring bases of DNA but are chemically modified to increase their likelihood of mispairing. The bases may be incorporated into DNA during its replication. 5-bromouracil, an analogue of uracil, can be incorporated into DNA in place of thymine. It can tautomerise to an enol structure that can pair with guanine (Figure 2.9) and, on replication, the basepair A-T will give rise to a G-C pair.

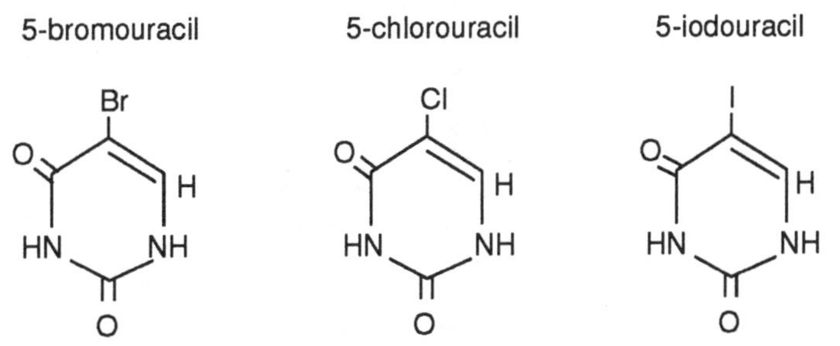

Figure 2.8 The structure of 5-bromouracil, 5-chlorouracil and 5-iodouracil. All three are in their keto form. Note that for simplicity, we have omitted the carbon atom from the ring structures.

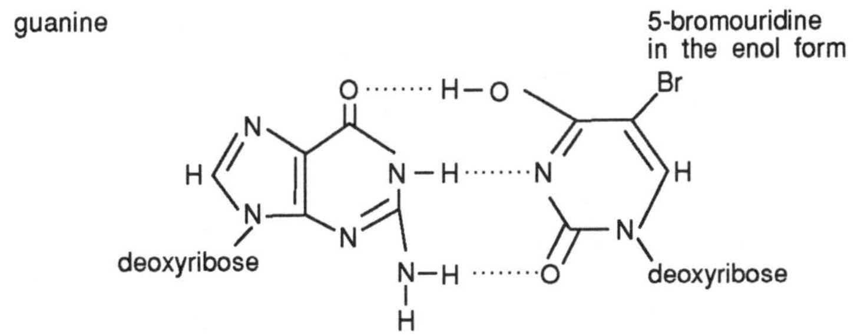

Figure 2.9 The enol form of 5-bromouridine pairs with guanine in DNA.

2-aminopurine (Figure 2.10a) is an analogue of adenine. It can pair with thymine or cytosine but in the latter case forms only a single hydrogen bond. The result is an A-T to G-C transition if it pairs with T or a G-C to A-T transition if it pairs with C (Figure 2.10b).

Other mutagenic base analogues are 8-azaguanine, and 6-thioguanine.

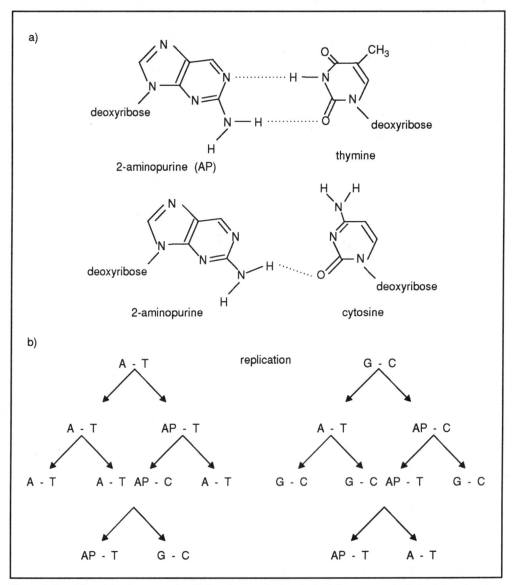

Figure 2.10 a) 2-aminipurine can pair with thymine or cytosine. In the latter case a single hydrogen bond forms. b) If 2-amino purine pairs with thymine, the result is an A-T to G-C transition or an G-C to A-T transition if it pairs with cytosine.

2.7.4 Chemical modification of bases

Bases in DNA can be specifically modified by chemicals to change their basepairing properties.

Nitrous acid and hydroxylamine

Nitrous acid deaminates bases. This may result in the change of A to hypoxanthine (which is read as G - Figure 2.11a), C to U which pairs with adenine (Figure 2.11b), and G to xanthine (which is read as C - Figure 2.11c). The net effect is an A-T to G-C transition or vice versa.

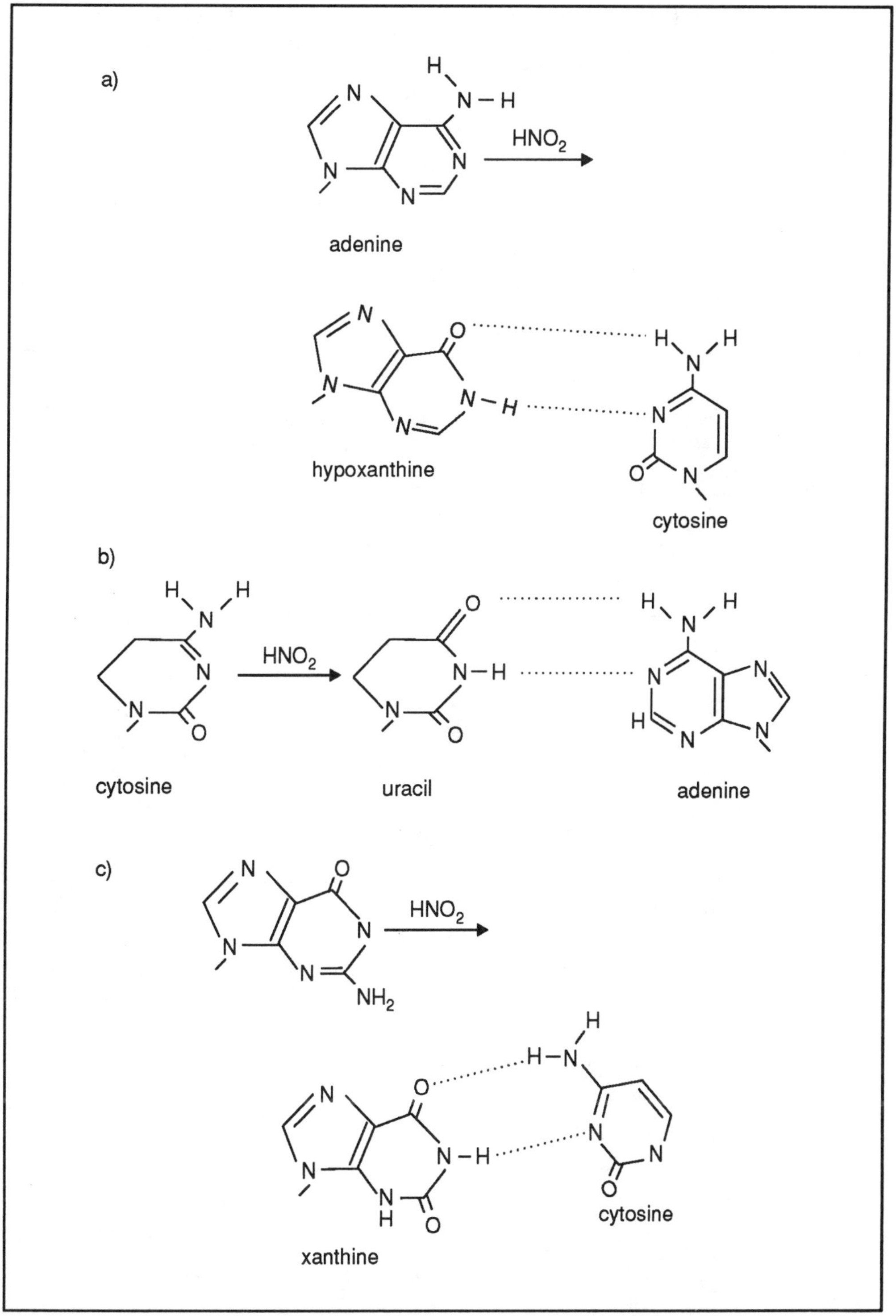

Figure 2.11 a) Nitrous acid deaminates adenine to hypoxanthine which pairs with uracil b) and cytosine to uracil which pairs with adenine and c) guanine is converted to xanthine which pairs with cytosine.

hydroxylamine

Hydroxylamine oxidises the amino group of cytosine to hydroxylaminocytosine (Figure 2.12) which produces similar changes in basepairings as nitrous acid. Thus a G-C to A-T transition will result.

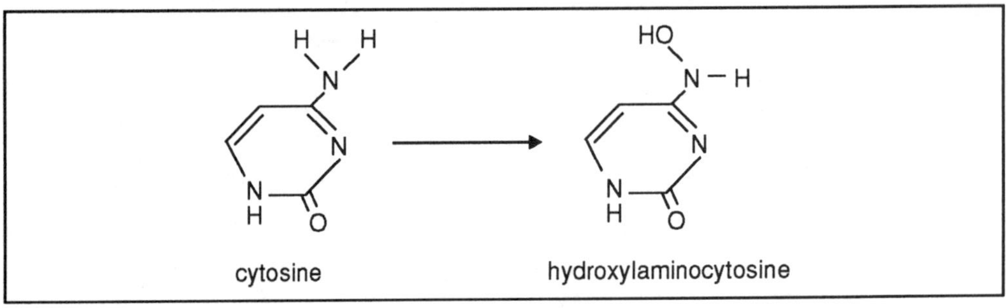

Figure 2.12 Hydroxylamine oxidises the amino group of cytosine generating hydroxylaminocytosine.

Alkylating agents

alkylating agents

Alkylating agents are compounds that produce a positively charged alkyl group, commonly CH_3^+ or $CH_3CH_2^+$, which react with, and chemically modify, bases in DNA. Guanine is especially susceptible to alkylation as it is exposed in the major groove of DNA. Alkylation occurs at positions N-7 and O-6 (Figure 2.13).

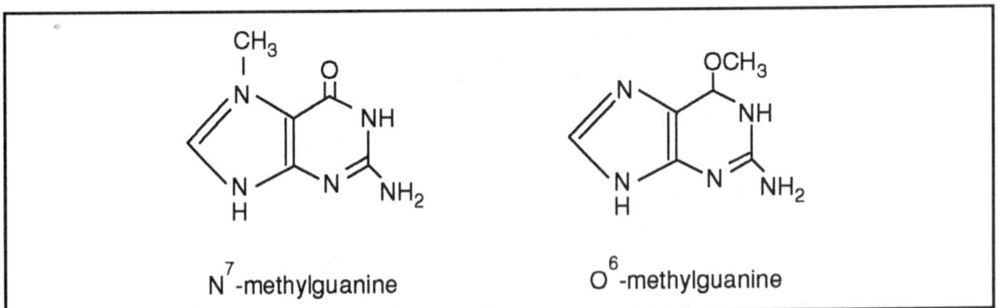

Figure 2.13 The structures of N7- and O6-methylguanine.

Alkylated guanine pairs with thymine rather than with cytosine and encourages a G-C to A-T transition. Position O-4 of thymine is also affected. Table 2.7 lists some alkylating agents.

Common name	Name
Mustard gas	di-(2-chloroethyl)sulphide
EMS	ethylmethane sulphate
EES	ethylethane sulphonate
MS	methylmethane sulphonate
DMS	dimethyl sulphate
MNNG	N-methyl-N-nitro-nitrosoguanidine

Table 2.7 Some commonly used mutagenic alkylating agents.

Dimethyl sulphate (DMS)and N-methyl-N-nitro-nitrosoguanidine (MNNG) are particularly effective alkylators. DNA bases may also be modified by spontaneous reactions in the absence of mutagens. Adenine can undergo hydrolytic deamination to hypoxanthine.

SAQ 2.9

Show how an A-T to G-C transition can occur after treatment of a section of DNA with the alkylating agent MNNG. We begin the sequence for you:

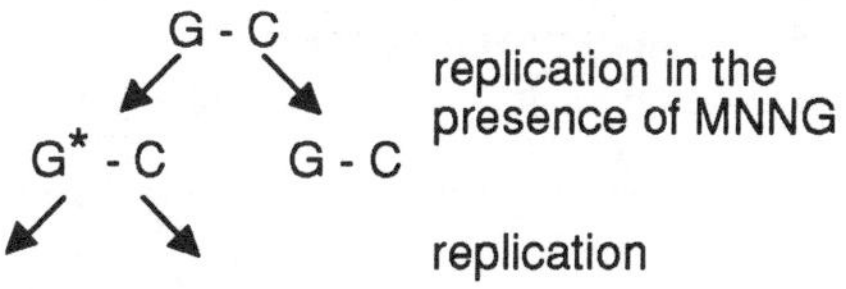

G - C

replication in the presence of MNNG

G* - C   G - C

replication

where $\overset{*}{G}$ represents alkylated guanine.

2.7.5 Intercalating agents

intercalating agents

Intercalating agents are planar molecules with dimensions similar to those of a basepair. Thus they are able to insert between basepairs in DNA leading to the addition or deletion of a nucleotide during replication. Common intercalating agents are ethidium bromide, proflavine and acroflavine (Figure 2.14). It is thought that intercalating agents lead to localised swelling and distortion of the double helix resulting in a frameshift mutation.

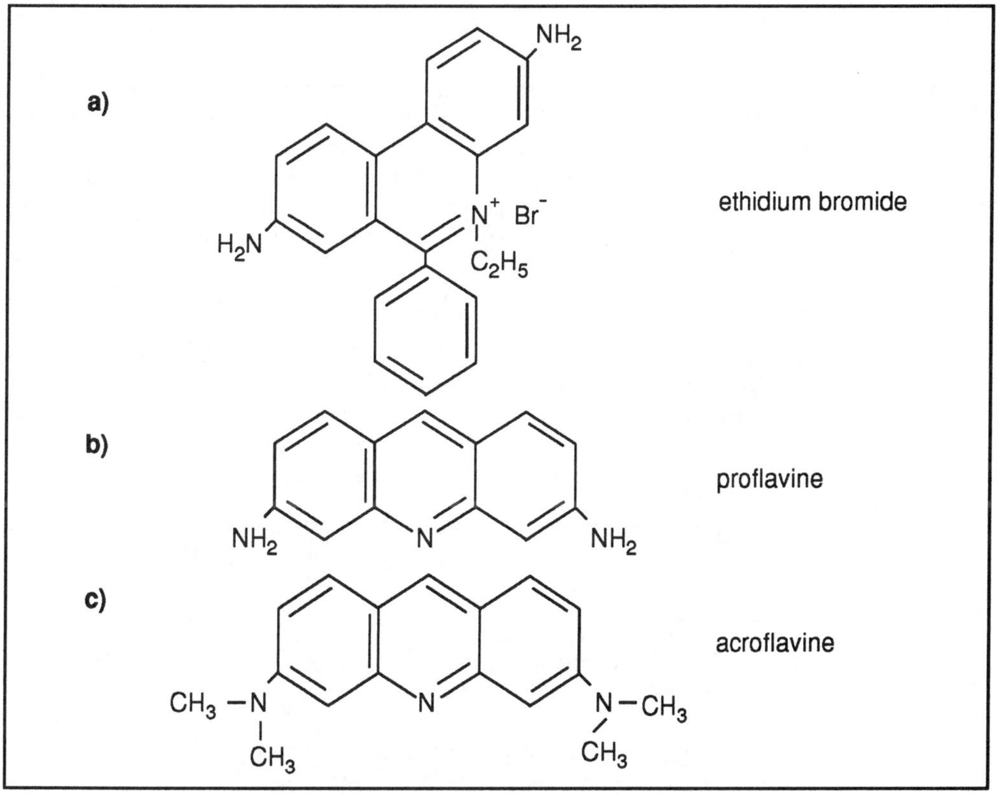

Figure 2.14 The structures of the intercalating agents. (See text for a description).

2.7.6 Radiation mutagenesis

thymine dimers

Radiation may damage sugar-phosphate bonds in DNA and cause breaks in one or both strands, chemically alter bases or cause cross-linking between DNA strands or between DNA and chromosomal proteins. The major defect is, however, the formation of thymine dimers which cause local distortion to the double helix.

Mutations can be induced in bacteria using either UV or X-rays. The source of mutation is the error-prone or SOS system which is inactivated allowing mutations to accumulate. Irradiation causes the production of thymidine dimers (Figure 2.15), commonly in the same strand of DNA, which may inhibit transcription of DNA.

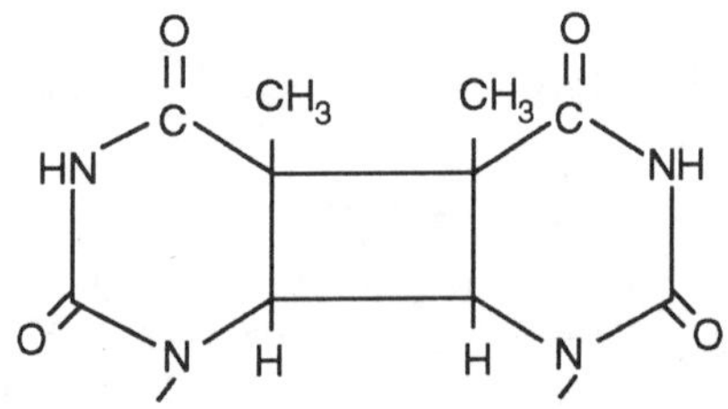

Figure 2.15 The structure of a thymidine dimer.

Survival curves

UV damage

Survival curves relate the proportion of surviving organisms in a culture to the radiation dose and can be used to determine the dose of radiation required to produce a measurable level of mutation. Double-stranded DNA is usually resistant to UV irradiation because the thymidine dimers can be excised and replaced. The undamaged DNA strand acts as a template and enzymes coded by the genes *uvr* A,B, C (*uvr* stands for UV resistant) are essential for the repair. If the bacterium is a mutant lacking *uvr* genes or genes coding for DNA pol I or III or DNA ligase it will be unable to repair UV damage. Such organisms will show increased susceptibility to UV irradiation.

Survival curves are shown in Figure 2.16. If a single interaction with UV light is sufficient to kill the organism then the proportion of surviving organisms (S) is given by the equation:

$$S = e^{Pn}$$

where n is the number of susceptible groups P is the probability that a group will be inactivated. This allows a second relationship to be derived:

$$P = KD$$

where K is a constant measuring the probability of unrepaired damage and D is the radiation dose.

Use Figure 2.16 to follow the description given.

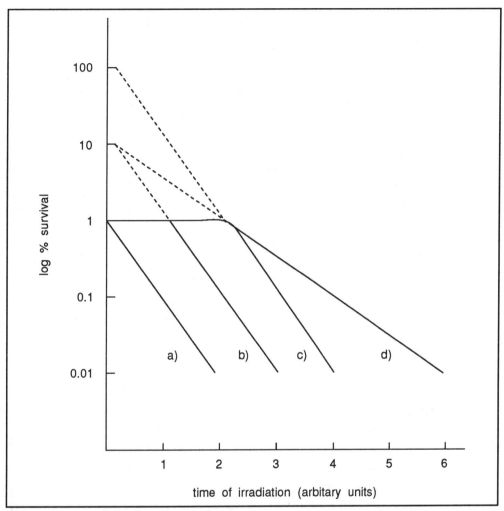

Figure 2.16 UV survival curves. The log % of survival is plotted against time of exposure at constant dose of radiation. The following curves are shown. a) An example of a single hit curve, b) a 10 hit curve and c) a 100 hit curve. a), b) and c) have the same slope indicating that they have the same genome size. Curve d) is also a 10 hit curve but the decreased slope indicates a smaller genome. The number of hits required to generate a mutant is obtained by extrapolation. Redrawn from B D Davies *et al*, 'Microbiology' (3rd edition),1980, Harper and Row, Maryland, USA.

The survival curve may be either a straight line, a) or a curve, b), c) and d). Curve a) is an example of a single-hit relationship where a single interaction with UV radiation alters the organisms viability. The slope of curve a) is -Krn (r is the radiation dose per unit time and n is the genome size). Curves b) - d) are multiple-hit curves where several interactions with radiation are necessary to affect the organisms survival. The number of interactions can be determined by extrapolating the curve giving 10 events for curve b) and 100 for c). Curves a), b) and c) have the same slope indicating the same genome size. d) Has an decreased slope indicating a smaller genome.

Survival curves can be used to predict the dose of radiation necessary to generate mutants. The dose-time product which gives about 0.01% survival is used.

2.8 Uses of mutations

Mutations have been widely used in basic research and also have uses in applied sciences and industry. For example, the inducement of mutations in experimental organisms have been widely used to identify individual steps in complex biochemical pathways (see Section 2.6).

Micro-organisms are used to produce a number of commercially important materials. However, the fermentation-based procedures do not usually rely on the wild-type organisms because they produce too low a yield of product. Rather, the organism is subjected to successive rounds of mutagenesis until a high yielding strain is produced. Table 2.8 shows the development of a modern high producing strain of a penicillin-producing organism from an original rather poorly yielding strain using a number of different mutagenic treatments and some lucky spontaneous mutations.

Strain of *Penicillium chrysogenum*	Yield (mg ml^{-1})	Treatment	Institution
NRRL-1951	60	spontaneous	Northern Regional Laboratory
NRRL-1951.B25	150	X-rays	Carnegie Institute
X-1612	300	ultraviolet light	University of Wisconsin
WIS Q-176	550	ultraviolet light	University of Wisconsin
WIS B 13-D 10	-	spontaneous	University of Wisconsin
WIS 47-638	-	spontaneous	University of Wisconsin
WIS 47-1564	-	spontaneous	University of Wisconsin
WIS-48-701	-	nitrogen mustard	
WIS 49-133	-	spontaneous	
WIS 51-20	-	ultraviolet light	Eli Lilly & Co
E-1	-	nitrogen mustard	Eli Lilly & Co
E-3	-	nitrogen mustard	Eli Lilly & Co
E-4	-	nitrogen mustard	Eli Lilly & Co
E-4	-	nitrogen mustard	Eli Lilly & Co
E-6	-	nitrogen mustard	Eli Lilly & Co
E-8	-	nitrogen mustard	Eli Lilly & Co
E-9	-	nitrogen mustard	Eli Lilly & Co
E-10	-	nitrogen mustard	Eli Lilly & Co
E-12	-	nitrogen mustard	Eli Lilly & Co
E-13		nitrogen mustard	Eli Lilly & Co
E-14		nitrogen mustard	Eli Lilly & Co
E-15	-	spontaneous	Eli Lilly & Co
E-15.1	7 g l^{-1}		Eli Lilly & Co

Table 2.8 The use of mutations in the development of high penicillin producing strains of *Penicillium chrysogenum*. Adapted from Primrose S B (1987) Modern Biotechnology, Blackwell Scientific, Oxford.

Table 2.9 summarises the effect of various mutation procedures on mutant production in *Neurospora*.

Treatment	Exposure time (minutes)	% Survival	Number of mutants/10^6 survivors
Spontaneous rate	-	100	0.4
Aminopurine	during growth	100	3
Ethylmethane sulphate	90	56	25
Nitrous acid	160	23	128
X-irradiation	18	16	259
Methylmethane sulphate	300	26	350
UV-irradiation	6	18	375
Nitrosoguanidine	240	65	1500
ICR-170*	480	28	2287

Table 2.9 The effect of various mutagens on mutant production in *Neurospora*. Taken from Sukuki *et al*, 'An introduction to Genetic Analysis' (1989) Freeman & Co, NY. * ICR -170 is an intercalating agent.

SAQ 2.10

Calculate the percentage increase in yield between the initial and final strains of the penicillin-producing organisms using the data in Table 2.8.

2.9 Delayed expression of mutants

After induction of a mutation with one of the agents described above the mutation may not be expressed until the organism has undergone some growth and several cell divisions. There may be several reasons for this. Bacteria may have more than one copy of the chromosome depending on the stage of the cell cycle. A mutation will probably only occur in one strand of one chromosome and it will take several cell divisions for all descendants of the mutated cell to acquire the mutation. If the mutation leads to a loss of a protein the bacterium is said to have a negative phenotype; the positive phenotype will continue to be expressed until all the non-mutated DNA is lost. This segregation lag   type of delay is called a segregation lag.

∏ Consider a bacterium which is allowed to replicate six times. 1) if a mutation arises in generation two what proportion of the offspring in generation six will be mutants? 2) If the mutation occurs in generation five what proportion of the offspring in generation six will be mutants?

The answer to 1) is one half of the offspring will carry the mutation. If you did not come to this conclusion, then draw a diagram like this:

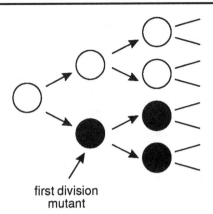

first division
mutant

Now continue until six generations are complete. You will find that half the offspring will carry the mutation.

The answer to 2) is 2 out of 32 offspring in generation six will carry the mutation.

phenotypic lag

Another reason for delay in the appearance of the mutation is that the functional protein may have a long life in the cell and may persist after the cell has mutated. Thus phenotypic lag will persist until all the protein has disappeared from the cell.

DNA is double-stranded and it is probable that a mutation will only arise in one strand. When proteins are synthesised an mRNA copy of one strand is made. If this is the non-mutated DNA strand then no phenotypic effect will be seen until the DNA has replicated.

2.10 Suppression of mutations

back mutation

reversion

Mutants sometimes re-express the characteristics of the wild-type by acquiring a second mutation. There are many ways in which a second mutation can neutralise the effect of the first. The simplest is a second mutation restoring the original nucleotide sequence of the DNA. This is called a back mutation and is a rare event. A reversion restores the original amino acid sequence coded by the gene but not the original base sequence.

Summary and objectives

Mutations are changes in the genome of a cell. The mutation may lead to both phenotypic and genotypic change and is the basis of biological variation. In micro-organisms the growth of the organism on a minimal medium is affected by the mutation. This altered growth pattern can be used to isolate mutants. Mutations arise spontaneously and can be induced using physical and chemical means. It is possible to a limited extent to mutate micro-organisms to increase their usefulness.

After reading this chapter you should be able to:

- recognise that mutations are the basis of biological variation;

- explain the types of mutation that can occur in DNA;

- show that mutations arise due to alteration in the base sequence of DNA;

- describe the methods of isolating mutants;

- know how to generate mutants using a variety of physical and chemical means;

- explain the effect of radiation on DNA and the use of survival curves to produce mutants;

- outline the use of mutants in research and industry.

Genetic recombination - transformation and conjugation

Genetic recombination - transformation and conjugation

3.1 Introduction

In Chapters 1 and 2 of this book we have studied the genetic organisation of the prokaryotic genome at the molecular level, looking at DNA structure, replication and the consequences of mutation.

In this and the next chapter, we will examine the use of micro-organisms for genetic research by considering *in vivo* genetic analysis, recombination and exchange. This study will be split into two. In this chapter we are going to study bacterial genetic systems. In the next chapter we will examine transposons and viruses which infect bacteria (bacteriophages).

3.2 Bacteria and their nature

Bacteria are simple, free living, unicellular organisms. They have a single, circular chromosome, which is not enclosed in a membrane. Their simple structural organisation is generalised and shown in Figure 3.1. Note that the chromosome is shown greatly folded in the figure. This is in fact how it occurs in bacterial cells. (See Chapter 1).

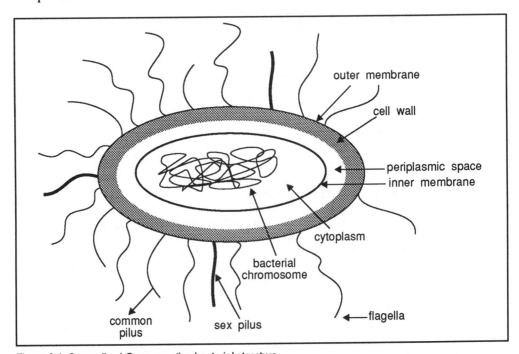

Figure 3.1 Generalised Gram negative bacterial structure.

Gram positive
and Gram
negative
bacteria

Bacteria come in several different shapes depending on their cell wall structure. There are two distinct classes of cell walls and the two may be distinguished by their ability to retain a stain called crystal violet when treated with alcohol. Bacteria which retain this stain are known as Gram positive, those which do not as Gram negative. The bacterium which we will be studying in this chapter is the rod shaped Gram negative bacterium *Escherichia coli* (abbreviated to *E. coli*). It is a well studied, commonly used bacterium and is normally found in the human gut. We are going to examine its genetic organisation with respect to recombination and exchange.

Prior to this we must introduce the basic information we need to understand the organism and how it grows.

3.3 Bacterial growth

Bacteria can be grown on a variety of media and compared to multicellular organisms their requirements are relatively simple. Growth occurs by division of one cell into two (binary fission) and is rapid due to the short life cycle (20 - 40 minutes).

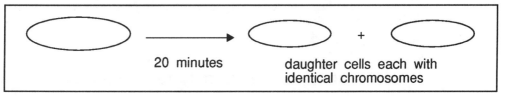

Figure 3.2 Division of *E. coli*.

SAQ 3.1

Assume that one bacterium finds its way into a container holding a liquid which is ideally suited to its growth. After approximately 7 hours it might have been able to divide 21 times. How may bacteria might be found there?

Ring the correct answer:

1) 1;

2) 18;

3) 19;

4) 36;

5) 900-1 000;

6) >1 000 000.

bacterial
colonies

A single viable bacterium can easily be detected if it is grown on a solid (agar solidified) nutrient medium. It will grow and divide to form a colony of many thousands of genetically identical individuals. (Figure 3.3)

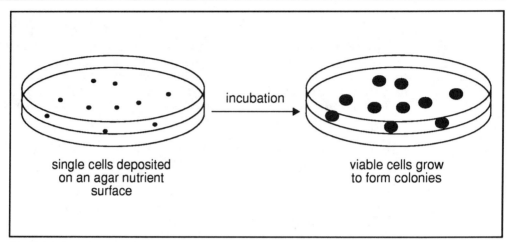

Figure 3.3 A nutrient agar plate containing bacterial colonies.

The ability of bacteria to grow and divide under various culture conditions is the major phenotypic (physical appearance in a particular environment) criterion used in bacterial studies. This contrasts with the term genotype which is used to describe the genetic make up of an organism.

3.4 The importance of bacterial mutants in genetic studies

As we learnt in Chapter 2, selective techniques are commonly available to identify and select mutant organisms, and are essential in studies of bacterial genetics. These techniques have allowed us to construct and use improved strains in industries such as brewing, chemical synthesis and pollution control and to understand genetic regulation. The techniques have also enabled us to genetically manipulate bacteria to produce proteins of interest such as insulin and various interferons.

In order to study genetic recombination and exchange it is important to have bacterial mutants with which we can follow such events. Because of the importance of the strategies we can use to isolate mutants, we will expand on the information we gave you in Chapter 2 on this topic.

3.5 Selection and types of mutants

Mutants may be selected by:

• sensitivity to chemicals;

• requirement for certain compounds for growth;

• ability to use (breakdown) compounds.

3.6 Sensitivity to chemicals

antibiotics

Bacteria are sensitive to (unable to grow in) the presence of some chemicals - often antibiotics such as streptomycin (Sm) are used. Sensitive organisms may be designated as - Sm S, resistant organisms - Sm R. Other antibiotics commonly used are tetracycline (Tet), kanamycin (Km), and penicillin and its derivatives.

3.7 Requirement for inclusion of specific nutrients in the culture medium.

Different bacteria have diverse nutritional requirements, and different media have been developed in which to grow them in the laboratory. All bacteria, however, require an energy source, a carbon source, nitrogen, sulphur, phosphorus, several metallic ions and water. Bacteria are usually grown on a chemically defined or synthetic medium. A medium which only supplies the minimum necessities required by the bacterial species being grown is called a minimal medium (MM). Many bacteria can grow on this very simple minimal medium in the absence of any organic substances other than a carbon source such as glucose, lactose or galactose. They are able to manufacture for themselves most complex compounds needed for growth such as amino acids, vitamins and lipids. Such an organism is said to be a prototroph.

prototroph

auxotroph

Other bacteria have a dependence on specific nutrients being included in the culture medium, as they cannot make all essential compounds themselves. An auxotroph is an organism which requires the presence of an organic compound (not as carbon source). A bacterium requiring the amino acid proline (pro) is a proline auxotroph, represented as pro^- and may be distinguished from the 'normal' (or wild type) strain shown as pro^+. Other commonly used auxotrophic markers are met^-, leu^-, his^-, trp^- and thr^- (strains requiring: methionine, leucine, histidine, tryptophan and threonine respectively).

3.8 Ability to use substances as nutrient sources.

As we have seen some bacteria are unable to make all the compounds they need. Others are unable to use (breakdown) certain nutrients. For example the inability to use lactose or galactose is represented as lac^- or gal^- the wild type being lac^+ and gal^+.

3.9 Indirect selection by replica plating

A rapid screening technique called replica plating has been developed to study nutritional requirements for bacterial growth. Bacteria are inoculated onto a Petri dish of nutrient rich agar-solidified medium so that a number of colonies are formed.

A piece of sterile velvet is then pressed gently on top of the plate. A few cells from each colony are transferred to the velvet, which is then pressed in a similar manner onto a new Petri dish this time containing a medium which lacks a specific nutrient. After time has been allowed for growth, the pattern of colonies found on subsequent plates will be in the same spatial orientation as that of the first dish except that colonies of bacteria which are auxotrophic for the missing nutrient will be present on the first plate but not on the later plate, see Figure 3.4.

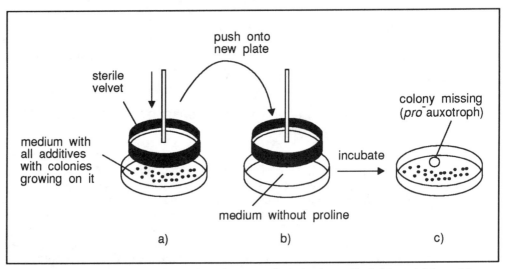

Figure 3.4 The technique for replica plating. a) A pattern of colonies from a Petri dish containing a rich medium is transferred; b) to a second plate lacking proline; c) where colonies fail to grow we can assume that the original colony in that location was a proline auxotroph.

If we wish to use that colony in further experiments we can go back to the original (master) plate remove a small number of cells and allow them to propagate in fresh proline containing medium.

SAQ 3.2

A typical minimal medium (MM) contains Na^+, K^+, Mg^{2+}, Ca^{2+}, Fe^{2+}, NH_4^+, Cl^-, SO_4^{2-}, phosphate buffered to neutrality (around pH 7) and a carbon source. In this case the carbon source is lactose. Note that the enzyme β-galactosidase hydrolyses lactose to produce galactose and glucose. The growth or otherwise of organism A on a variety of agar plates containing minimal medium and some supplements are reported in Table 3.1 Which of the following five descriptions best describe organism A?

1) his^-, pro^-, auxotroph, can breakdown lactose and is SmS;

2) his^-, pro^-, prototroph, can breakdown lactose and is SmS;

3) his^-, pro^-, auxotroph, requires Sm for growth, can breakdown lactose;

4) his^+, pro^+, prototroph, can breakdown lactose and is SmS;

5) his^+, pro^+, auxotroph, can breakdown lactose and is SmS.

Medium/Supplement	Growth +/-
MM + lactose + proline	-
MM + lactose + histidine	-
MM + lactose + histidine + proline	+
MM + lactose	-
MM + lactose + histidine + proline + Streptomycin	-

Table 3.1 Growth of organism A in minimal medium (MM) with supplements.

From Table 3.2 below, we can see organism B is a *trp*⁻ auxotroph. Its inability to grow on MM + lactose unless glucose is added shows that it is unable to use (breakdown) lactose as a carbon source.

Medium/Supplement	Growth +/-
MM + lactose	-
MM + tryptophan + lactose	-
MM + lactose + glucose	-
MM + tryptophan + lactose + glucose	+

Table 3.2 Growth characteristics of organism B.

If we were to mix organism A (described in SAQ 3.2) with organism B, under the right circumstances we would produce individuals capable of growth on MM containing lactose. This could be due to the transfer of genetic information between the two and is a common event in some bacterial species.

3.10 Natural transfer of genetic information in bacteria

transfer of
genetic
information

Bacteria have several different ways of transferring genetic information. These mechanisms may be divided into 4 main categories, (which as we can see later, can themselves be subdivided):

- **transformation** - uptake of naked DNA into recipient bacteria;

- **conjugation** - DNA is transferred from a donor (male) to a recipient (female) via a specialised sex pilus;

- **transposition** - movement of DNA from one position, to another within the genome, due to small mobile genetic elements called transposons;

- **transduction** - bacterial genes are carried from a donor to a recipient via a bacteriophage.

The first two of these headings are discussed in this chapter. Phage genetics, transduction and transposition are discussed in Chapter 4.

3.11 Bacterial transformation *in vivo.*

episome

Transformation is the process in which recipient cells take up extraneous DNA from the environment and incorporate it into their genomic DNA through recombination. A piece of DNA which integrates into the chromosome and is maintained and replicated in this state may be referred to as an episome. Such DNA pieces can arise naturally (from dead, lysed bacteria), are double-stranded and are generally large, on average 20,000 nucleotide (base) pairs long. Small pieces may also be taken up, but a minimum length of about 500 basepairs is required in order for integration into the host chromosome. As we shall see, without recombination and integration into the host chromosome (unless a piece of DNA can maintain and replicate itself - see Section 3.12.2) it will be lost and not transmitted to the progeny.

competent cells

Many bacterial species such as *E. coli* are capable of such uptake. Cells capable of DNA uptake are said to be competent. Not all cells show such competence and there are wide variations in the efficiency of transformation between bacterial species.

Transformation was once an important means of determining gene order in microbial genetics and remains still, for some organisms, the only means of mapping (ordering) genes. For example genes from species such as *Bacillus subtilus* have been studied extensively through the process of transformation. Transformation in *E. coli* is however less efficient, although this organism does possess very efficient conjugation and transduction systems.

Both the number of competent cells present in a culture and the length of time that a cell may remain competent is variable. The process is active (energy requiring) and to be transformed the cell must posses surface proteins which bind DNA.

In *E. coli* the main genetic information (a few thousand genes) is in the bacterial chromosome which is composed of a circular double-stranded DNA molecule. It is highly condensed (folded and coiled). This DNA is packed into the central region of the cell and is not bounded by a nuclear membrane.

3.11.1 The mechanism of transformation

The natural transformation process may be split into several stages. The main ones are:

- reversible binding of double-stranded DNA molecules to cell surface receptor sites;

- irreversible uptake of donor DNA into those bacterial cells that are competent;

- conversion of donor DNA into single-stranded molecules via degradation of one strand;

- integration of all or part of the single-stranded donor DNA into the recipient chromosome;

- segregation and phenotypic expression of donor DNA in the recipient.

homologous recombination

There are many different types of recombination, but let us assume, that the introduced DNA has large regions of homology with part of the bacterial chromosome and that DNA exchange is brought about by the general (or homologous) recombination system.

This system is responsible for the synapsis (interaction between the two homologous DNA strands). As we show in Figure 3.5 double-stranded DNA is converted to single-stranded DNA upon entry into the bacterial cell.

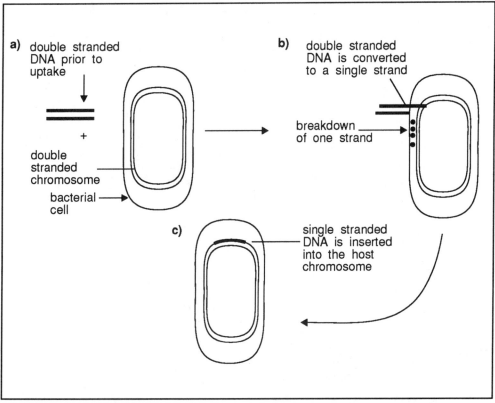

a) double stranded DNA prior to uptake

+

double stranded chromosome

bacterial cell

b) double stranded DNA is converted to a single strand

breakdown of one strand

c) single stranded DNA is inserted into the host chromosome

Figure 3.5 Bacterial transformation. Double-stranded donor DNA is converted into a single-strand which is inserted into the host bacterium's chromosome by recombination. For simplicity the bacterial chromosome is shown as a double-stranded circle, although in reality it is highly coiled, folded and compact.

The recombinational events are shown in more detail in Figure 3.6 and are as follows:

RecA protein a) single-stranded DNA once in the bacterial cell is coated with RecA protein, to form a nucleoprotein. RecA protein is a protein coded for by the so called *rec A* gene. It is a gene that was first discovered as being essential for certain types of transformation. Its name was derived from the fact that it was involved in genetic recombination;

b) and c)
displace-
ment loop
Rec A protein together with other factors appears to unwind the double-stranded bacterial chromosome forming what is known as a displacement loop (D loop). The invading strand moves into this loop and pairs with the region of homology;

d) chiasma (or cross-over) occurs, then breakage and reunion, the gaps are filled in and the strands are rejoined, incorporating the new strand, instead of the old;

e) finally any mismatches are repaired.

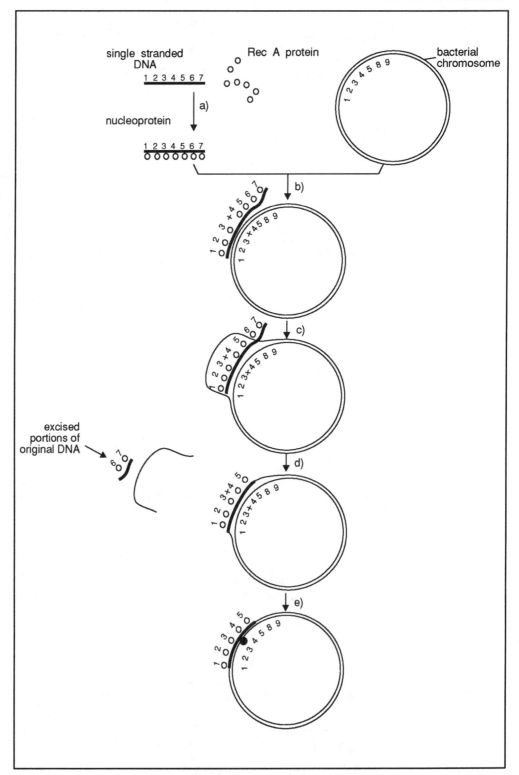

Figure 3.6 Recombination in bacterial transformation. 1, 2, 3, 4 , etc signify areas of homology; + signifies an area of mismatching: ● signifies a region in which mismatching has been corrected (see text for details).

This homologous recombination requires regions of extensive homology and is promoted by the chromosomal *rec A* gene protein product (Rec A). In *rec A⁻* bacteria this process will not occur.

non-homologous recombination

Non-homologous recombination in contrast does not require a marked sequence similarity, or the *rec A* gene product. Non-homologous recombination may be split into two categories; site specific and illegitimate recombination. These will be dealt with in a later section.

3.11.2 Use of transformation for gene mapping

gene mapping

Transformation has often been used as a technique for gene mapping (ordering genes) in the bacterial chromosome. To illustrate this let us consider the two genes *leu* and *his* in *E. coli* strains Y and Z. The genotype of strain Y, which is able to make leucine and histidine for its self will be written *leu⁺, his⁺*. Strain Z is an autotroph of genotype *leu⁻, his⁻*. Let us assume for the sake of argument that these genes are on opposite sides of the chromosome. When DNA is isolated from the bacterium the chromosome will shear into smaller bits and the two genes will be on physically separate pieces of DNA (Figure 3.7).

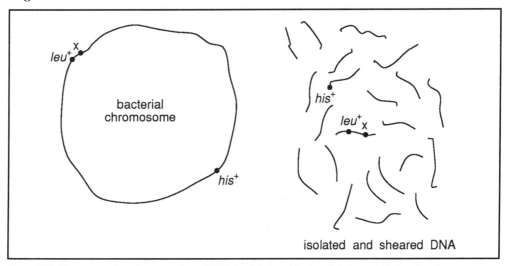

Figure 3.7 The separation of genes when DNA is isolated and broken into fragments by shearing.

A very small amount of this DNA may be put in the culture medium of *E. coli* strain Z under conditions that favour transformation. Assume that some of the bacteria (perhaps 1%) are competent. After a time interval to allow transformation to take place, transformants are simply selected by their ability to grow in the absence of either or both of these amino acids.

As transformation is a relatively rare event it is unlikely that a particular recipient that is transformed for the *leu* gene will also be transformed for the *his* gene as this would require two transformation events!

Let us now consider in addition a gene x that is situated directly next to the *leu* gene position, (= locus). It is highly likely that during DNA isolation these two genes will remain together on the same piece of DNA (see Figure 3.7). Therefore there is a good possibility that if a recipient is transformed with one trait it will also be co-transformed with the second trait, associated with the adjacent genetic locus. Thus if two genetic markers are closely linked, double transformants may be found at a frequency approaching the frequency of single marker transformants in a comparable single marker experiment.

So we can see that the frequency with which two markers are co-transformed can be used as a crude estimate of the distance between them.

SAQ 3.3

Which of the following experiments would you use to determine if *leu* and *his* loci were close together on the bacterial chromosome.

Circle the correct response.

1) mix DNA from *leu*⁻, *his*⁻ bacteria with *leu*⁺, *his*⁺ bacteria and plate onto a medium lacking both leucine and histidine.

2) mix DNA from *leu*⁺, *his*⁺ bacteria with *leu*⁻, *his*⁻ bacteria and plate onto a medium lacking both leucine and histidine.

3) mix DNA from *leu*⁺, *his*⁺ bacteria with *leu*⁻, *his*⁻ bacteria and plate onto three media one lacking leucine, another lacking histidine, and a third lacking both.

If instead we wished to do a contrasting experiment to obtain a set of auxotrophic mutants we could do the experiment in the reverse manner. DNA from the auxotrophic *leu*⁻, *his*⁻ bacterial strain could be used to transform prototrophic *leu*⁺, *his*⁺ bacteria. This time the selection process would be more laborious as the numbers of auxotrophic transformants are low in number compared to the nontransformed prototrophic bacteria. We are unable to select auxotrophic bacteria directly by their growth on a particular medium but must select them indirectly following the replica plating technique discussed in Section 3.9. Many hundreds of bacteria from rich histidine and leucine containing medium would have to be replica plated onto media lacking either or both of these amino acids. Auxotrophs could be identified indirectly by their failure to grow in media lacking the required amino acids.

3.11.3 The use of penicillin in indirect selection of transformants.

If all we want are the auxotrophic recombinants from the above experiment it is possible to eliminate the untransformed cells. This can be done using an adaption of the penicillin enrichment procedure we described for isolating mutants in Chapter 2. Let us remind you of the processes involved. Firstly growth conditions are employed which only allow the growth of untransformed bacteria (such as the use of a liquid minimal medium without leucine or histidine). Secondly the antibiotic penicillin can be added. Penicillin acts by preventing the final stage of cell wall synthesis and will kill all growing bacteria (in the absence of a normal cell wall the bacteria will burst). In contrast non-growing bacteria which are not attempting to build a cell wall will be unaffected. Finally the penicillin can be washed away and those cells which have not been killed *penicillin* (those auxotrophs previously unable to grow) are placed onto Petri dishes containing *enrichment* minimal medium supplemented with leucine and histidine and are allowed to grow. This procedure increases the ratio of auxotrophs to wild-type cells in the suspension.

∏ An experiment using very small amounts of transforming DNA shows that the following genes are co-transformed.

a) U and T   c) P and Q

b) Q and R   d) R, T and S

What is the probable gene order?

Write down your answer on a piece of paper before moving on.

Their sequence can be found from the following information. For two genes to transform together they must be close together. Therefore we can tell that U and T must be near neighbours as are Q and R and P and Q. Q must be between R and P, as P and R do not co-transform. S must be between T and R because R, T and S co-transform.

The order therefore is: U; T; S; R; Q; P (or P; Q; R; S; T; U).

Information such as relative gene position is very important when we are considering recombinational events that rearrange genes or events such as conjugation (Section 3.13) in which genes are transferred from a fixed point in a specified direction from one organism to another.

Not all genes have easily identifiable products and it may be difficult or impossible to rapidly select for the product they encode. In such a situation we can use the presence of a closely linked gene. Let us say that gene Q is of interest but we cannot directly select for this gene. Let us also assume that we knew that the product of a gene R was closely linked to Q and that the product of gene R makes cells resistant to streptomycin (SmR). Similarly on the other side of the gene of interest, there is a gene P, the product of which makes cells resistant to tetracycline. If we transform a SmS, TetS bacterial host with DNA and select cells that are SmR and TetR we can assume with a high probability that our non-selectable gene of interest (gene Q) will also be there.

Many Gram positive bacteria lyse and release DNA naturally on ageing, or in various deleterious conditions. This, together with the process of transformation, has probably been important during the evolutionary process.

3.11.4 *In vitro* transformation

It is possible in the laboratory to induce very high transformation efficiencies, by mixing bacteria with DNA and then treating them with complex solutions of chemicals, or by subjecting them to high, short duration electrical pulses. In such circumstances the transformation process is not thought to be by the same mechanism. We will not deal with these processes here. These, and other processes of transformation, are dealt with in the BIOTOL text, 'Strategies for Engineering Organisms'.

3.12 Plasmids and conjugation

3.12.1 Plasmids

plasmids

So far we have considered the bacterial chromosome as the main source of genetic information. Bacteria often also contain one or more genetic elements called plasmids. These plasmids may contain important genes, which although dispensable by the host in most growth conditions have particular value in certain environments. For example conferring on the host the ability to degrade an environmental pollutant such as toluene or to grow in the presence of antibiotics such as streptomycin. These genes are only of benefit to the host organism in the presence of such compounds but have little or no effect on the bacterium in the 'normal' environment.

3.12.2 General properties of plasmids

replicon

Plasmids are double-stranded circular supercoiled DNA molecules and are relatively small (0.2 - 4% of the size of the bacterial chromosome). By definition a plasmid is a replicon, (unit of genetic material capable of replication independently from the chromosome), that is stably inherited without being integrated into the bacterial chromosome. Again in our studies of plasmids and conjugation we will use *E. coli* as this organism has been comprehensively studied and is well understood.

3.12.3 *E. coli* plasmids

antibiotic
resistant
colicins

E. coli plasmids can be split into three main groups: F, R and Col types. In this Chapter we will predominantly focus onto the F-type the so called sex or fertility factor. Cells carrying the F plasmid (F$^+$) are able to transfer genes to a recipient (F$^-$) via a sex pilus. F$^-$ strains do not carry an F factor. R plasmids are generally less easily transferred. R plasmids carry antibiotic resistance genes. Col plasmids code for species-specific bacteriocidal proteins called colicins.

3.12.4 The F factor

The F factor, is of particular interest to us as it can exist both as a plasmid (autonomous) and as an episome. An episome integrates into the bacterial chromosome and replicates along with it. We shall see the importance of this later.

The F factor can mediate its self-transfer from cells with the plasmid to cells without the plasmid, via a sex pilus in a process called conjugation which involves DNA transmission between cells.

A donor cell with an F factor in an autonomous state is described as F$^+$ (male), upon conjugation with a recipient F$^-$ (female) only the F factor is transferred and both exconjugants (cells involved in conjugation) become F$^+$. Transfer can occur once or twice per generation. Thus by mixing a population of F$^+$ cells with F$^-$ virtually all cells in the new population quickly become F$^+$.

3.12.5 Incompatibility groups

plasmid
replication

Although many plasmids use host machinery to replicate themselves there are many significant differences in the mechanisms and enzymology of plasmid replication.

copy number

stringent
control

Each type of plasmid possesses its own genes for controlling its rate of replication initiation and hence copy number per cell. We may classify plasmids on this basis. Stringent control - low copy number plasmids - are those that occur at only 1 or 2 copies per cell. The partitioning of such plasmids at cell division is generally controlled to avoid total plasmid loss from the cell.

relaxed control

Relaxed plasmids are present at a high copy number of 10-100 per cell. Thus if a cell is transformed with one copy of a high copy number plasmid, the plasmid will be replicated rapidly to reach a high number. This process is thought to be controlled by a plasmid encoded repressor that negatively regulates initiation of replication, ie the more plasmid present the more repressor present which limits its further replication. Repressor activity may depend on concentration. So when a cell enlarges before dividing into two the repressor concentration drops and replication occurs until twice the number of plasmids are present, at which time sufficient repressor is present to prevent further replication. For high copy number plasmids a greater concentration of repressor is necessary to limit replication than for low copy number plasmids.

∏   What would happen if a cell contained two types of plasmids, one of low copy number and under stringent control, the other of high copy number under relaxed control if they were both controlled by the same repressor?

We would expect to lose the low copy number plasmid as the culture grows. The production of repressor as a consequence of the presence of the relaxed plasmid would switch off the production of the low copy number plasmid. The average number of these low copy number plasmids per cell would therefore progressively fall as the cells grew and divided. Many of the progeny cells would, therefore, fail to receive a copy of such plasmids.

Thus plasmids with similar mechanisms of replicative control (for example, controlled by the same repressor) are generally not compatible in the same cell and only one of them will be stably maintained.

incompatibility groups

Hundreds of plasmids have been sorted out into Incompatibility or Inc groups. Incompatibility is tested by determining the ability of different plasmids to cohabit in a cell. Up to 7 different plasmid types may cohabit individual cells. Plasmids from the same Inc group cannot coexist together in a cell.

3.12.6 Barriers to transfer

entry exclusion

As we have seen incompatibility operates once the plasmids are in the same cell. Another mechanism can operate before this stage which actually prevents the plasmid getting into the cell. This process is known as entry (or surface) exclusion. It involves products from the two, F coded genes, *tra S* and *tra T*. The proteins coded for by these genes are responsible for reduced pair formation between bacteria carrying homologous plasmids and operate at the level of the cell envelope.

| SAQ 3.4 | As we have introduced a lot of new terms up to this point we will pause briefly so that you can see how well you have assimilated this information. Read through the terms 1) - 12) and match them by writing them on a piece of paper with the letter of the appropriate definition. |

Terms		**Definitions**	
1)	episome	a)	donor bacterium which contains an F factor
2)	phenotype	b)	genetic constitution
3)	R plasmid	c)	organism which can grow on a simple medium without a requirement for additional complex organic compounds except as carbon source
4)	genotype	d)	observed characteristics of an organism in a particular environment
5)	auxotroph	e)	able to take up naked DNA from the surrounding medium
6)	competent bacterium	f)	transfer of donor genes to a recipient via a sex pilus
7)	conjugation	g)	protein involved in generalised recombination
8)	Rec A protein	h)	organism which cannot grow on minimal medium but which requires complex organic compounds other than for carbon source
9)	F^+	i)	recipient bacterium which lacks a F plasmid
10)	prototroph	j)	drug resistance plasmid
11)	F^-	k)	circular piece of DNA which is able to maintain itself and replicate independently of the chromosome
12)	plasmid	l)	piece of DNA which integrates into the chromosome and is maintained and replicated in this state

3.13 Conjugation

Cells that carry the F plasmid are morphologically distinct from F⁻cells. In addition to the large numbers of small common pili frequently present, there are also a few (typically 1 to 3 for a rapidly growing culture) long thin proteinaceous appendages called F (or sex) pili. They take the form of long hollow tubes (see Figure 3.1). F pili are said to be male specific as they disappear if the F plasmid is lost. The F pilus is the site of adsorption for 'male specific' bacteriophages (bacterial viruses), these phages can therefore be used as probes for the presence or absence of F pili. Conjugation is the process of transfer of DNA from a donor to a recipient cell and is assumed to occur through a sex pilus, which is often known as a conjugation tube. This process does not occur naturally in many species, although it is common in *E. coli*. Conjugation is one way of transfer of genetic information. It is an important means of *in vivo* genetic manipulation as useful genes may be moved from one host to another.

3.13.1 Conjugation, DNA transfer and replication

We may summarise the conjugation process in *E. coli*, using the F plasmid as an example:

Formation of donor/recipient pairs following effective cell/cell contact

Formation of the mating pair appears to occur after a random collision, which is dependent on cell density, temperature and media viscosity. Close cell/cell contact is thereafter achieved by the use of the F pili. They bring about initial contact and then draw cells into close contact.

R pili whose synthesis is controlled by the R plasmid, can also mediate such events.

Remember that cells may contain more than one type of plasmid.

Conjugative (transmissible) plasmids may provide conjugation facilities for non-conjugative plasmids which would not on their own mediate DNA transfer. They may be used experimentally to allow movement of an otherwise non-transmissible plasmid from one host to another. Such plasmids are known as drivers or mobilising plasmids.

Preparation of DNA for transfer (mobilisation)

Some plasmids such as the F plasmid are able to prepare their own DNA for transfer. An F plasmid encoded endonuclease (the *tra* YZ gene product) makes a single-stranded nick (cuts the sugar phosphate chain) in the transfer origin (*ori* T) of the plasmid. This mobilisation is usually plasmid specific.

DNA transfer

The 5′ end of the nicked strand is then transferred through the pilus into the recipient, leaving the complementary strand behind in the donor. Both strands are then immediately used as templates for the synthesis of double-stranded DNA by bacterial DNA polymerase. This form of replication is called rolling circle replication.

After DNA replication is complete both cells contain an F plasmid (see Figure 3.8).

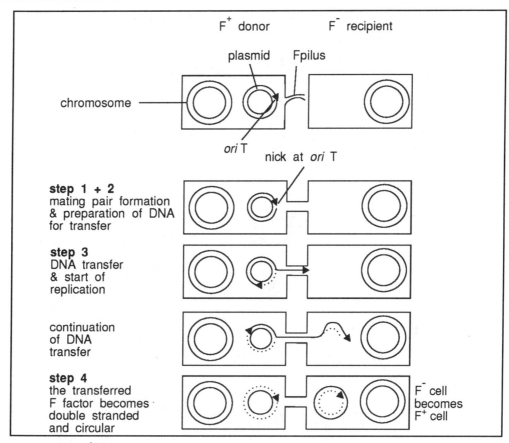

Figure 3.8 The F⁺/F⁻ conjugation process. (See text for further details).

3.13.2 F mediated gene transfer

An F factor can integrate into the host chromosome at one of many sites and in either orientation. Integration is thought to occur following a reciprocal crossover event to form a single circular structure (see Figure 3.9a) and requires a functional host recombination system (*rec* A⁺). Integration sites are not random, they show homology with the F plasmid and integration is thought to be mediated by insertion (IS) elements (we will describe them in more detail later in this text). A cell containing an integrated F factor is known as Hfr, which stands for high frequency recombination - the reason why will soon become apparent.

3.13.3 The role of the Hfr in chromosomal DNA transfer

bacterial
chromosome
mobilisation

Bacterial chromosome mobilisation is similar to DNA mobilisation during F transfer. In the Hfr strain, transfer is also initiated at the *ori* T. The F factor still mediates its self-transfer as we have seen previously, however, this time the attached chromosome may also follow into a recipient F⁻ cell (see Figure 3.9). Genes on the leading side of *ori* T (represented as 'bcd' in Figure 3.9b) will enter the recipient first, followed by chromosomal genes (in the order 3, 2, 1, 6, 5, 4). Only very rarely will the whole chromosome pass through the delicate conjugation tube as it often breaks and the cells separate before the entire transfer process is complete.

In Hfr x F⁻ matings, the recipient will usually remain F⁻during conjufgation as a portion of the integrated F factor (marked 'a' in Figure 3.9) is the last part of the genome to be transferred. Only occasionally will the entire F factor be successfully transferred in which case the recipient becomes Hfr. This is in contracts to $F^+ \times F^-$ crosses where the F^- will become F^+ (Figure 3.8). F^+

a)   The F factor (represented as the dark circle, with the four reference points a, b, c and d) integrates into the bacterial chromosome (the white circle with the numbers 1 - 6 representing different parts of the chromosome) to form an Hfr.
     Note the arrow head represents the first part of the DNA to be transferred during conjugation, see b) below.

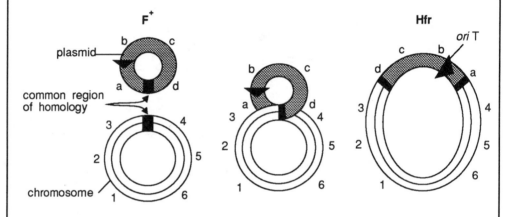

b)   One strand of Hfr begins to transfer into the F^- cell. Replication occurs via the rolling circle mechanism as in Figure 3.8.

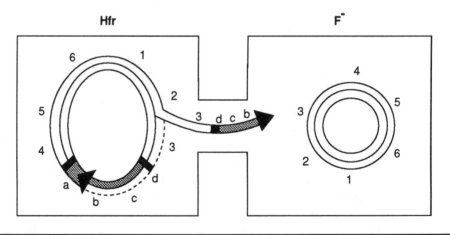

Figure 3.9 Hfr conjugation.

Chromosome transfer proceeds at a fairly steady rate (about 10^4 nucleotides per minute) and takes about 100 minutes to complete, this is in contrast to the 2-3 minutes taken by transfer of the F factor. The direction of chromosome mobilisation is governed by the orientation of *ori* T in the inserted plasmid and the site of insertion determines which chromosomal genes are transferred first.

In the same way as we showed in our discussion on transformation, markers which are close to each other are likely to be transferred together into the recipient cell.

recombination

The chromosomal part of the donor strand is very likely to have homology with part of the recipient chromosome and may integrate into it by recombination. The recipient chromosome is replaced by the transferred fragment through breakage and rejoining following recombination into the recipient chromosome. The integrated part of the donor fragment is therefore stably maintained and passes along with the rest of the chromosome to the daughter cells upon cell division. If the chromosomal segment is not integrated into the recipient chromosome it will not be stably maintained and will be lost on cell division.

Under optimum conditions, the transfer frequency of genes close to the leading edge approaches 100%. Once transferred these sequences can integrate following recombination at an efficiency of about 50%.

Figure 3.10 shows the process of gene transfer with time.

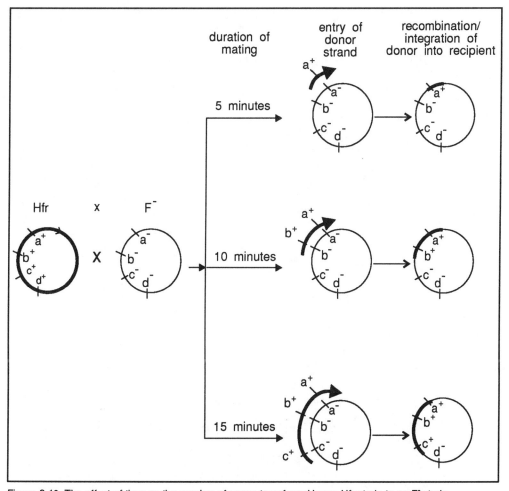

Figure 3.10 The effect of time on the number of genes transferred by an Hfr strain to an F⁻ strain.

The conjugation bridge is very delicate and may easily be broken. In the laboratory the mating can be stopped after different time intervals by violent agitation such as that produced by a blender, and the extent of transfer determined.

selection of recombinants

When carrying out genetic analysis, it is necessary to be able to identify the recipients and to be able to distinguish them from both of the parental types. The most common procedure is to use a recipient that has a recognisable genetic marker that is not present in the donor. This marker is located at such a position in the chromosome that it is likely that a mating pair of bacteria will have separated before the equivalent locus in the donor can be transferred. Antibiotic resistance markers such as streptomycin resistance are frequently chosen. The recipient in this case would be SmR, the donor Hfr SmS. A marker which is effective against the donor is known as the counter selected marker.

counter selected marker

After a period of mating, the cells will be plated on a medium which contains streptomycin. This will prevent the growth of Hfr cells but will allow both the F^- and the recombinant recipients to grow. The F^- parental and the recombinant recipients can then be distinguished from each other by additional selection using auxotrophic/prototrophic markers. For example if the donor is prototrophic trp^+ and the recipient trp^-, trp^+ recombinants can be selected by their growth on a medium from which tryptophan is omitted. The transferred marker, that is selected for, is known as the selected marker.

selected marker

Here are the details of a typical experiment.

A Hfr leu^+, trp^+, met^+, SmS donor was mixed with an F^- leu^-, trp^-, met^-, SmR recipient at time zero.

After the periods specified in Table 3.4 aliquots were removed and conjugation tubes disrupted using a Waring blender. Samples were placed on supplemented minimal media (MM). Glucose was included as carbon source, the additives were as in Table 3.3 which also shows the codes given to the media.

Medium MM		Code MM
MM + leu, trp, Sm	(no met)	-m
MM + trp, met, Sm	(no leu)	-l
MM + leu, met, Sm	(no trp)	-t

Table 3.3 Media codes used in the experiment described in the text.

Table 3.4 gives the number of bacterial colonies found on each type of plate (MM, -m, -l or -t).

Conjugation was interrupted after the times specified. The volume sampled was the same in each case.

Medium	Number of colonies												
	Time (minutes of conjugation)												
	0	2.5	5	7.5	10	12.5	15	20	25	30	35	40	45
MM	100	100	100	100	100	100	100	100	100	100	100	100	100
-m	0	0	0	5	11	18	25	29	30	29	30	29	30
-l	0	8	19	30	40	41	41	41	40	40	41	41	41
-t	0	0	0	0	0	0	0	0	0	7	15	19	20

Table 3.4 Results of the experiment described in the text. The results record the number of bacterial colonies produced after plating out cells which had been allowed to conjugate for varying lengths of time.

SAQ 3.5

Have a close look at Tables 3.3 and 3.4 and then ring the correct response True (T) or False (F) for the statements below.

1) Sm prevents the F⁻ from growing. T/F

2) MM lacks all of the amino acids (leu, trp, or met) required by the F⁻ and will not support its growth. T/F.

3) Neither parental type can grow on: -m, -l, or -t. T/F.

4) If the F⁻ recipient receives and has integrated into its chromosome the gene which encodes synthesis of the auxotrophically required amino acid leucine, it will be able to grow on the medium -l. T/F.

5) The gene encoding synthesis of leucine is transferred before that for tryptophan. T/F.

6) The selected markers are *met⁻*, *leu⁻* and *trp⁻*. T/F.

7) The counter selected marker is SmS. T/F.

If you had any problems with this you should re-read the section on selection (3.7 and SAQ 3.1) and that on conjugation (from the beginning of 3.13) as it is very important you understand the principles behind the selection against both donor and recipient which only allows recombinants to grow.

From the minimal medium results, we can work out the total number of Hfr cells/volume (as the Hfr can grow and the F⁻ cannot).

From the results with the other media we can work out the % Hfr cells that have donated DNA which includes the marker (leu, trp, or met) which is under consideration. Only such recombinants are capable of growth on -l, -t and -m which lack one of these amino acids but contain Sm. The data for each marker is shown in Figure 3.11.

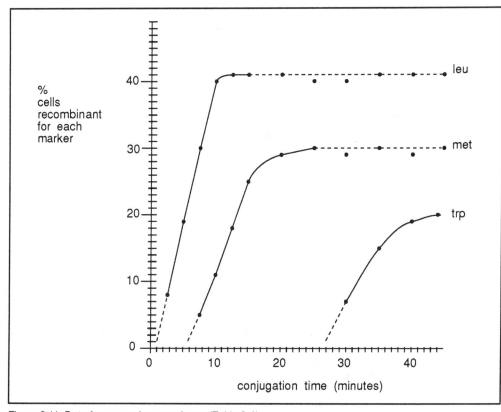

Figure 3.11 Data from a mating experiment (Table 3.4).

interrupted-
mating
experiments Such experiments are called interrupted-mating experiments and are frequently used
to map gene order by working out from the data to find the time at which the gene
under consideration entered the recipient.

Look at Figure 3.11. Each curve has a linear region which may be extrapolated back to
the time axis and used to give a value for the time at which each marker first enters the
bacterium.

You should see from the graph that there is a time before which no recombinants are
detected, this is because, as chromosome transfer precedes at a set rate, it takes a certain
time for the first marker to enter the recipient. The slow initial appearance of markers
is due to the fact that not all donor cells begin to transfer the DNA at the same time.

From the graph we can determine the times that the markers first appear as:

• leu : approximately 1 min.

• met : approximately 5.5 min.

• trp : 27 min.

In fact, apart from markers at the very beginning of the transfer or towards the end, the
time taken for genes to appear is a good indication of distance. A map distance of 1

minute corresponds to the length of chromosome transferred in 1 minute during conjugation.

We can use this information to construct a map of the chromosome.

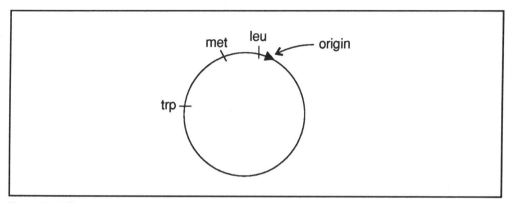

Figure 3.12 Chromosome map constructed from interrupted-mating experiment. (See text for description).

chromosome
map

The number of recombinants for each type reaches a plateau value, which is lower the later the marker enters the recipient (this is mainly due to the fragility of the conjugation tube).

Transfer begins at an *ori* T within the F factor which is integrated within the Hfr. Genes are transferred in order either clockwise or anticlockwise depending on the orientation of the F factor in the chromosome as shown in Figure 3.12.

A standard *E. coli* chromosome map is built up from a number of experiments using different Hfr strains each with the F factor inserted in a different position or orientation.

It is divided into minutes and starts at 0 (arbitrarily *thr* A, the threonine locus) continuing to 90 or 100.

3.13.4 The importance of recombination in conjugation

In the preceding sections we have seen how DNA is transferred into cells. It must undergo recombination in order for it to integrate into the host chromosome and be maintained through subsequent cell divisions. Without recombination (for example in a *rec⁻* host) we would not be able to see recombinant colonies following Hfr x F⁻ mating (Section 3.13.3). This is due to the fact that in bacteria, with the exception of a few phages, only circular DNA which contains a suitable replication origin is able to replicate and hence pass to future generations. This, we should note, is also true of DNA taken into the cell during transformation. In fact linear DNA is degraded by bacterial nucleases. In F⁺ x F⁻ matings, the situation is obviously different as it is likely the entire plasmid will be transferred and will therefore be able to maintain itself and replicate in its circular form.

In Hfr x F⁻ matings or following transformation with linear DNA, it is necessary to have two (or an even number) of recombination events flanking the gene(s) to be incorporated in order for the incoming DNA to be integrated into the circular chromosome, see Figure 3.13 a). With only one (or an odd number) of recombination events, see Figure 3.13 b), the DNA would loose its circular nature and be unable to replicate properly.

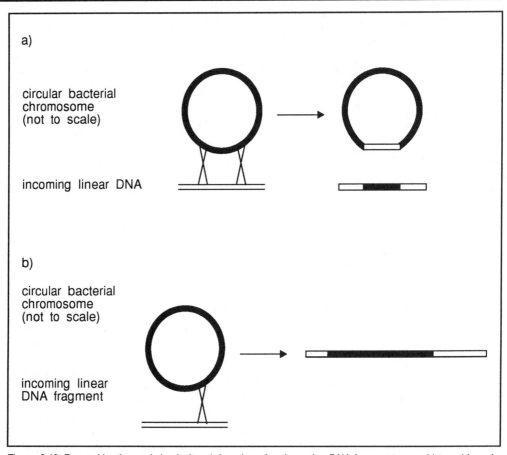

Figure 3.13 Recombination and circularity. a) A region of an incoming DNA fragment recombines with and integrates into, a homologous region of the bacterial chromosome following two recombinational events. Note the reciprocal DNA exchange. b) With only one recombination event the exchange seen in part a) does not occur. Instead the resulting product is no longer circular but linear and is unable to replicate itself!

Time of entry experiments such as those previously discussed give an overall picture of the *E. coli* genome but cannot give us much fine detail.

recombination frequency

The recombination frequency in *E. coli* is quite high and recombinations may occur several times between markers as little as a few minutes apart. Thus the linkage mapping scheme explained in the transformation section (Section 3.11.2) is not applicable to very large DNA segments.

In order to explain this, consider the following. Using time of entry experiments we have demonstrated a gene order:

leu, met, trp at 1.0, 5.5 and 27 minutes respectively.

In order to show close linkage we would assess colonies showing recombinant properties for one prototrophic marker gene for the presence of the others.

If just two recombination events occurred as in Figure 3.14, then the alleles *met*⁺ (5.5 mins), *X* (5.55 mins) and *trp*⁺ (27 mins) would be integrated into the F chromosome. However in reality further recombination events would occur similar to those shown shown in Figure 3.15.

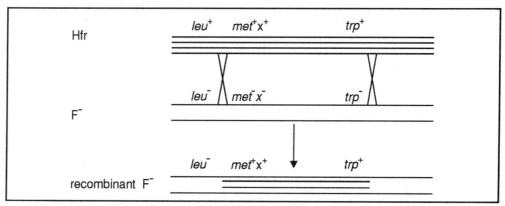

Figure 3.14 Two recombination events.

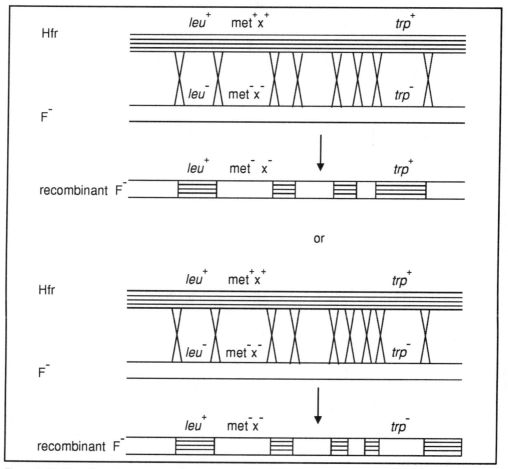

Figure 3.15 The effect of multiple recombination events.

As the distance between two genes increases the probability of a recombination event between them increases. Thus only genes very close together like *met* and *X* are unlikely to be subject to such events and are likely to stay linked together in the recipient.

At this stage it is important to remind you that recombination enzyme deficient bacteria are not used during strain construction processes that require recombination events.

Hfr x F⁻ matings have proven very useful in giving the relative order of genes that are not too close to each other as it is difficult to time their transfer into recipient cells accurately enough (see Figure 3.11). Phage transduction and two or three point crosses (discussed in Chapter 4) fill in a lot of the fine detail on the map as does sexduction which we will describe in Section 3.14.

SAQ 3.6

Three Hfr strains are mated individually with an auxotrophic F⁻ strain using the interrupted mating technique we discussed in Section 3.13.3. Using the data below order the genes in the *E. coli* chromosome drawn below. Assume it takes 100 minutes to transfer the entire chromosome. Note that we have already marked on the position of the *thr* locus.

Approximate time of entry

donor loci	HfrA	HfrB	HfrC
thr	22	96	62
his	66	52	6
thy	81	37	21
rha	8	10	48
lac	30	88	70
man	57	61	97

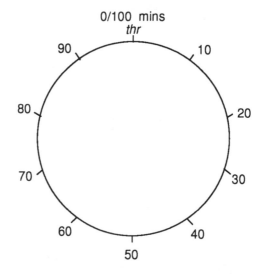

3.14 F′ mediated sexduction

The process of Hfr formation is reversible and it is possible, although more rare, for the F factor to be excised (cut out). Occasionally an error occurs and results in an F factor which contains both F and bacterial genes. The number of these genes excised varies from 1 gene to half of the chromosome. This hybrid is termed an F prime (F′). Such incorrect excisions are thought to be caused by faulty synapsis (cross overs) between regions of partial homology, rather than between the fully homologous correct sequences (the terminal F sequences). It is possible to get two different sorts of abberrant excision and this is illustrated diagrammatically in Figure 3.16.

abberrant
excision

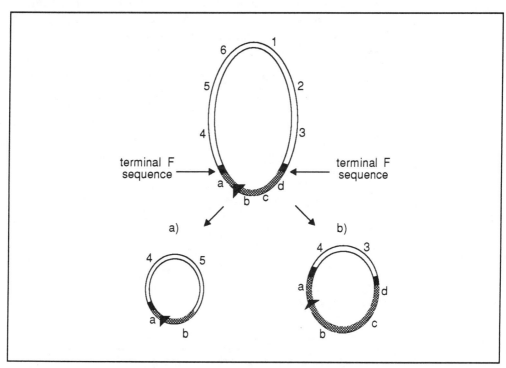

Figure 3.16 Faulty excision of the F plasmid.

In the first type of aberrant excision (see Figure 3.16 a)) reciprocal recombination occurs between a region on the bacterial chromosome and one within the integrated F plasmid. In this instance a substituted F′ is produced which has lost some of the F DNA but gained some bacterial DNA. This F′ will only be stably maintained if the F region which is involved in vegetative replication is still present. Moreover if essential *tra* genes are lost the plasmid will no longer be able to induce conjugation. All of these points must obviously impose constraints on the type of aberrant excision events which can be detected.

In the second type of aberrant exchange (see Figure 3.16 b)), recombination occurs between regions of the bacterial chromosome which border the F DNA. The F plasmid is therefore intact but carries with it additional bacterial DNA from either side of the insertion point. The plasmid may contain very large portions of the bacterial chromosome, in some cases up to half!

Either type of plasmid is usually essential to the survival of their host, depending on the number of genes involved. In their absence the bacteria could lose a significant amount of DNA.

3.14.1 Transfer of F', into a recipient

sexduction   Transfer of F', into a recipient occurs in F'/F⁻ matings. This process is called sexduction.

As the F' factor replicates autonomously the F⁻ becomes F', and establishes a new line of F' cells. This transconjugant line will be partially diploid (merodiploid), if, as is likely, it carries its own copy of the region of the bacterial chromosome carried by the F', element (see Figure 3.17).

Homology between the bacterial DNA on the F' plasmid and the merodiploid bacterium may lead (in a *rec*⁺ strain) to recombination and an exchange of genes. Alternatively the F' plasmid may insert itself into the host chromosome via a region of homology.

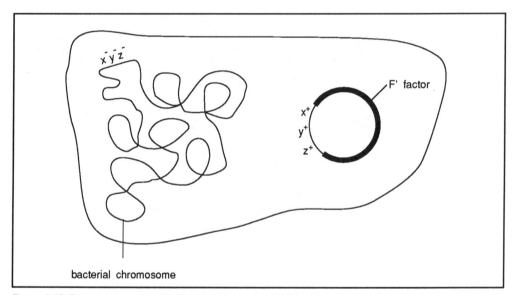

Figure 3.17 F' containing cells partially diploid (merodiploid) for x y z.

Partial diploids such as these are very useful in the study of dominant/recessive relationships at genetic loci and can give useful information on gene interactions. As bacteria normally have a single copy of a particular gene such analysis would otherwise be difficult.

SAQ 3.7

You are provided with a SmS Hfr culture of bacteria as shown below. An interrupted conjugation experiment gave the following times for transfer of genes.

a$^+$ 10 minutes

b$^+$ 14 minutes

c$^+$ 25 minutes

d$^+$ 83 minutes

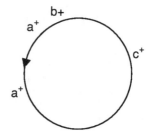

This strain is mated with a *rec*$^-$, a$^-$, b$^-$ c$^-$, d$^-$ SmR F$^-$ recipient culture. Five colonies result after selection on minimal medium containing added a, b, c and streptomycin but lacking d.

Which of the following may be true? Ring true answers.

1) The 5 colonies are d$^+$ prototrophs.

2) The Hfr has conjugated with the F$^-$, transferred the d$^+$ gene which has integrated into the recipient chromosome.

3) The resistance to streptomycin comes from the Hfr strain.

4) The Hfr strain may have contained some F′ plasmids one of which containing the d$^+$ allele has been transferred.

5) In order for the d$^+$ gene to be expressed it must integrate into the F$^-$ chromosome.

3.15 Alternative strategies for studying the prokaryotic genome.

In this chapter we have learnt that the *E. coli* genome is readily manipulated. Transformation and conjugation have proven to be valuable tools in the study of prokaryotic genomes. However, not all bacteria are as amenable to such techniques as *E. coli* has proven to be. In the following chapter we will examine the viruses which infect bacteria and study the structure and function of mobile genetic elements called transposons. We will in addition consider in more detail the use of recombination and two or three point crosses in genetic mapping.

Summary and objectives

In this Chapter we have examined the processes of genetic recombination in prokaryotes. We began by discussing the importance of bacterial mutants in genetic studies before examining bacterial transformation. We then went on to discuss plasmids and conjugation with particular reference to the F factor and Hfr strains. We also studied how genetic recombination between strains enables us to map the genome of prokaryotes such as E.*coli*.

Now you have completed this chapter you should be able to:

- understand the following terms: phenotype, genotype, prototroph, auxotroph, wild type, plasmid, episome, replica plating, transformation, conjugation and recombination;

- appreciate the existence of different types of bacterial mutants and understand their uses in *in vivo* manipulations;

- understand the difference between direct and indirect selection;

- be able to contrast F, R and Col plasmids;

- be able to state two possible mechanisms behind the inability of two plasmids to be found in the same bacterial cell;

- be able to explain the *in vivo* transformation mechanism in *E. coli*;

- understand how to use co-transformation experiments to determine gene order;

- be able to explain the role of the F factor in conjugation between F^+/F^-, Hfr/F^- and F'/F^- bacteria and explain the role of interrupted mating in experiments to determine gene order;

- show diagramatically how we can obtain two types of F' plasmid.

Transduction and transposition

Transduction and transposition

4.1 Introduction

In Chapter 3 we have looked at transformation and conjugation, and have seen how recombinational events play an important role in these processes.

bacteriophages

insertion
sequences

transposons

However, in addition to plasmids and the DNA comprising the chromosome, bacteria such as *E. coli* may contain at least three other DNA species. These are bacteriophages, insertion sequences and transposons. In this chapter we will consider these 'new' DNA element with respect to their genetic organisation and their ability to insert donor DNA into recipient molecules.

4.2 What is a bacteriophage?

bacteriophages

In brief, bacteriophages, also known as phages, are bacterial viruses. Their structure is relatively simple. They comprise a molecule of DNA (sometimes RNA) which is packaged (or encapsulated) into a protein coat, the capsid. The chromosome carries a variable number of genes, depending on the type of phage. Phage replication can only occur inside a suitable host cell.

The structure of a typical phage, for example phage lambda, is shown in Figure 4.1.

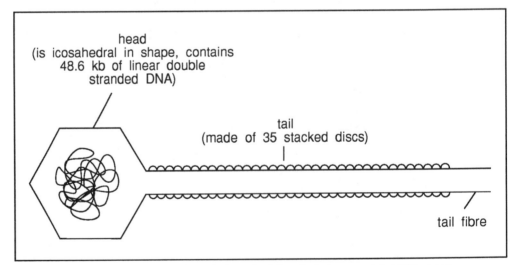

Figure 4.1 Stylised representation of bacteriophage lambda, a typical encapsulated DNA phage.

4.3 Bacteriophage life cycles

There are two alternative life cycles by which phages reproduce:

• the lysogenic life cycle involves a stage known as lysogeny, during which the phage genome lytic functions are repressed.

• the lytic life cycle in which the phage replicates, matures and lyses the host cell releasing the phage progeny.

lytic, lysogenic and temperate phages

We will discuss these in detail in the following sections. Some phages, for example T4, are virulent. That is, when they infect a sensitive bacterial cell they reproduce and eventually lyse the host cell, releasing many progeny phage. In other words they are in a lytic life cycle. Phage lambda (often written as λ) has a more complicated life style and may exist either in the lytic or lysogenic mode. Bacteriophages which have both types of life style are known as temperate.

∏ Write down the reasons why it might be of benefit to a phage to be capable of both types of life style?

Which pathway is followed depends on a complex interaction of both genetic and environmental factors. The lytic pathway, for example is often favoured when the bacterial host is growing in a nutrient rich environment. The lysogenic pathway predominates when conditions are poorer and enables the phage to survive until the environment is once again favourable. More than 90% of the thousands of known phages are temperate.

In order to understand the process of transduction (transfer of bacterial genes from a donor bacterium to a recipient via a bacteriophage) we will briefly consider each life cycle in turn using phage lambda as our example.

4.4 The lambda life cycle

Adsorption

The lambda infective process begins with the tail fibre (see Figure 4.1) recognising and binding to a special receptor site on the bacterial cell surface. This adsorption requires the gene product of the *E. coli lam* B⁺ gene. The tail then penetrates the cell and the λ DNA is injected into the bacterium, whilst the empty capsid remains on the outside.

cos sites

Once within the cell the linear λ DNA circularises by pairing between complementary bases in its two, twelve bases long, cohesive sticky ends, known as *cos* sites (see Figure 4.2).

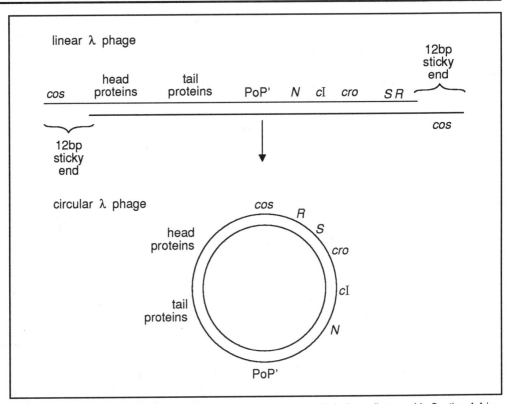

Figure 4.2 Circularisation of lambda DNA. The functions of the genetic loci are discussed in Section 4.4 in relation to the lytic and lysogenic life cycles.

At this stage several genes, known as the early genes, are expressed and the phage will then either enter the lysogenic or lytic pathways.

Lysogeny

lysogeny

During the λ lysogenic life cycle instead of replicating rapidly, the phage DNA integrates into the host cell chromosome, this is known as lysogeny. Integration is via a reciprocal recombination event occurring between a special attachment site on the circularised λ chromosome (*att*P) and a corresponding but not identical site on the bacterial chromosome (*att*B). Each attachment site comprises a 15 basepair core sequence common to all *att* sites and represented as (o) (cross over occurs between two (o) regions). On either side of the (o) core region are a pair of flanking, dissimilar sequences: B and B' on the bacterial chromosome and P and P' on the phage chromosome. The site *att*B may be represented as BoB' and *att*P as PoP'. Integration generates two new hybrid attachment sites BoP' and PoB' (see Figure 4.3).

integrase

This event is important as it represents a special type of recombination - site specific recombination. It is *rec* A$^+$ independent and can occur in *rec* A$^-$ hosts. Site specific recombination is entirely dependent on the λ *int*$^+$ gene product, a protein known as integrase.

As the bacterial cell divides the phage genome behaves as is it were a group of bacterial genes and replicates synchronously with the bacterial chromosome passing along with it into the daughter cells. In its integrated state the phage genome is known as a prophage. The frequency of lysogeny varies from zero to nearly 100%.

prophage

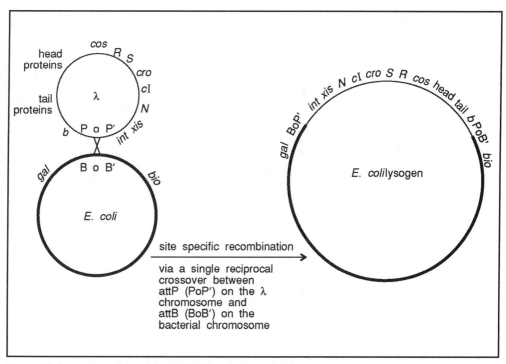

Figure 4.3 Integration of circurlaised λ DNAinto the bacterial chromosome.

SAQ 4.1

Write down whether the following statements are true or false.

1) Bacteriophages are unable to replicate in the absence of a suitable host cell.

2) The nucleic acid in the λ phage head is double-stranded linear DNA.

3) In the λ lysogenic life style, free phages are released following replication and breaking open of the host cell.

4) Upon infection of a bacterial cell by a λ phage the following sequence of events occurs: adsorption of phage to the bacterial cell, circularisation of phage DNA within the phage head, injection of phage DNA into the bacterium.

5) The attachment site on the circular λ chromosome is known as *att* B and that on the host chromosome as *att* P.

6) Site specific recombination is *rec* A⁺ dependent.

SAQ 4.2

If the gene order inside the head of phage λ is as below:

cos A B C PoP' *D E F cos*.

What will the prophage gene order be in the bacterial host during lysogeny?

lysogenic state

The lysogenic state is maintained due to the *cI* gene product which remains active during lysogeny. The cI gene product is a repressor which is transcribed from the promoter for repressor maintenance (*pRM*). It acts by binding to two prophage operator-promoter sites known as *oLpL* and *oRpR* and inhibits the expression of the other lambda genes (see Figure 4.4), so preventing replication and cell lysis. For this reason a lysogenic cell is immune to secondary infections by the same virus, as the lytic genes of the incoming virus would also be repressed by the *cI* repressor.

Transition from the lysogenic to the lytic state occurs naturally, although rarely (1 in 10^5 phage). It also may be induced by, for example, irradiation with ultra violet light. Induction in this instance occurs because the UV indirectly destroys the *cI* repressor.

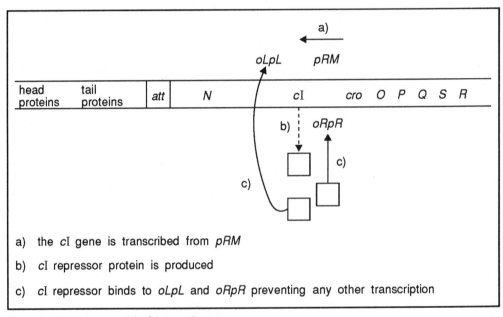

a)   the *cI* gene is transcribed from *pRM*

b)   *cI* repressor protein is produced

c)   *cI* repressor binds to *oLpL* and *oRpR* preventing any other transcription

Figure 4.4 Maintenance of the λ lysogenic state.

The lytic life cycle

Occasionally the prophage can leave the bacterial chromosome, either spontaneously or by induction. The prophage excises by the reverse of the integration process (see Figure 4.5). This occurs via site specific recombination between the hybrid attachment sites BoP' and PoB' and requires not only the integrase protein but also the product of

excisionase

the λ*xis⁺* gene otherwise known as excisionase.

lytic state

In the lytic state, in the absence of the *cI* repressor, RNA polymerase is able to bind to *oLpL* and *oRpR* and transcribes the genes known as the early genes. The *N* and *cro* gene products are made. These allow transcription of the delayed early genes and replication of lambda DNA. After completion of DNA replication, the late genes are switched on. These encode the structural components of the head, tail, and tail fibres and other proteins which are involved in the assembly of these components and their packaging into the phage heads. Finally the products of genes S and R cause the host cell to lyse, releasing about 100 phage progeny from the cell (see Figures 4.6 and 4.7).

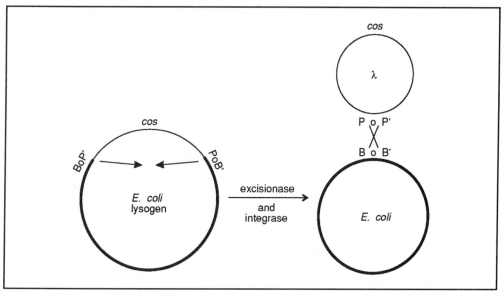

Figure 4.5 Excision of λ from the bacterial chromosome.

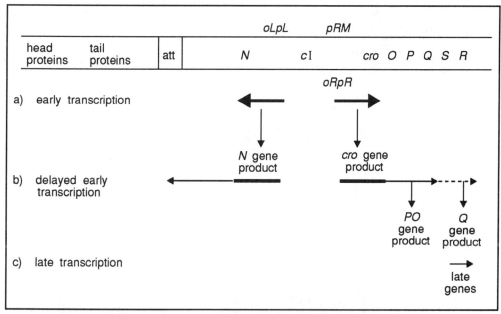

Figure 4.6 Gene expression during the λ lytic life cycle. a) RNA polymerase binds to *oLpL* and *oRpR* and transcribes the early genes. b) The *N* gene product allows transcription of the delayed early transcription. c) The *cro* gene product prevents further delayed early transcription. The *O* and *P* gene products allow DNA replication. The Q gene product promotes late transcription.

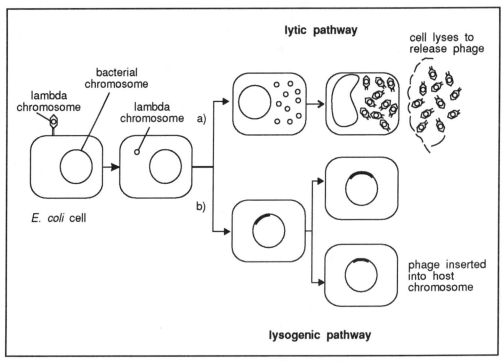

Figure 4.7 The λ life cycle. a) In the lytic life cycle phage DNA replicates independently from the bacterial chromosome, phage products are made, progeny phage assembled and released upon cell lysis. b) In the lysogenic life cycle the phage DNA integrates into the bacterial chromosome and replicates along with it.

We can see the result of the lytic process. If healthy bacteria are grown on an agar plate they form an opaque layer of growing cells. Where bacterial lysis and phage release occurs, a clear area known as a plaque is formed.

plaque

SAQ 4.3

Write down whether the following statements are true or false.

1) The λ lysogenic state is maintained due to the *cI* gene product.

2) The *cI* gene product is an inducer.

3) UV light may initiate the change from lysogeny to lysis.

4) Temperate phages only have a lytic life cycle.

5) Excision of the λ prophage requires only the product of the λ *xis⁺* gene (excisionase).

6) The sequence of events during the λ lytic cycle may be briefly summarised in the following order: RNA polymerase binds to *oLpL* and *oRpR* and transcribes the early genes. The *N* and *cro* genes allow the transcription of the delayed early genes and λ DNA replication occurs. The late genes are switched on and allow synthesis of phage structural components and their assembly and packaging. The *S* and *R* genes cause the host cell to lyse releasing phage progeny.

4.5 Transduction

Having looked at phage life cycles we can consider the means by which phages bring about transduction. Transduction is the process by which phages transfer bacterial DNA from a donor bacterium to a recipient cell.

Two types of transducing phages are known:

- generalised;

- specialised.

generalised transducing phages

Generalised transducing phages produce particles that contain only DNA obtained from the host bacterium rather than phage DNA. This DNA may be derived from any part of the bacterial chromosome.

specialised transducing phages

Specialised transducing phages produce particles that contain both phage and bacterial DNA linked in a single molecule. The bacterial genes are obtained from a particular region of the bacterial chromosome.

Π The transduction mechanism depends on how the phage grows in bacterial cells. By pausing here and considering the two types of life cycle can you think why this is so? Note your thoughts down on a piece of paper and we will discuss the mechanisms in the following sections.

Temperate phages are capable of specialised transduction, examples of such phages include lambda and φ80. The chromosome of such temperate phages integrates at only one, or more rarely a few specific attachment sites within the host chromosome.

4.6 Production of specialised transducing particles

The prophage excision process is generally precise. Occasionally (1 in 10^6), however, a mistake is made. Host genes close to the site of insertion may be excised with phage DNA and packaged into phage particles. This will occur only if the spacing between the 2 cut sites gives rise to a molecule 76-100% of the length of the normal lambda phage genome, due to constraints on phage lambda packaging.

illegitimate cross-over

The incorrect excision process is shown in Figure 4.8. Illegitimate cross-over occurs between non-homologous regions of the phage and the *E. coli* chromosome either to the right or to the left of the prophage, leading to the specialised transducing particle carrying either the *E. coli bio* or *gal* genes. The formation of λ*gal* and λ*bio* transducing particles entails loss of λ genes. The number of missing genes depends on the position of the cuts. In our example λ*bio* lacks the *int* and *xis* genes and is able to form plaques. This particle is therefore written as λp*bio* (the p denotes its plaque forming ability. In other cases where the deletion extends into essential *N* genes the phage is defective and cannot form plaques and is written λd*bio*).

Head and tail genes are essential, so for example *gal* transducing particles which lack these genes are defective and cannot form plaques and may be written λd*gal*.

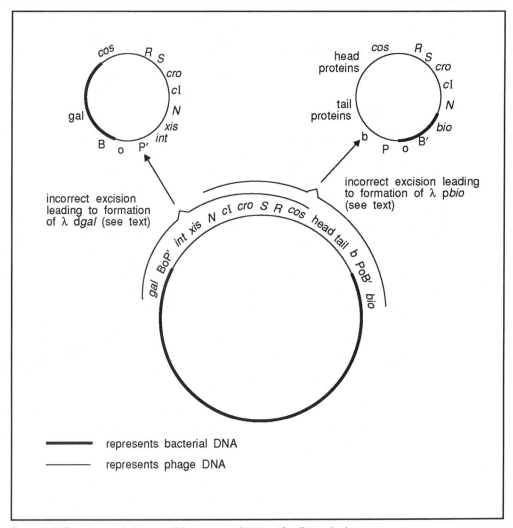

Figure 4.8 The abnormal phage excision process (see text for discussion).

You may find the following two SAQs difficult to answer, if this is the case read through Sections 4.4 and 4.5 again.

| SAQ 4.4 | Explain how it is that λdgal transducing phages can be produced when we know the phage lacks the genes for the head and tail. |

| SAQ 4.5 | Explain why λpbio phages which lack the xis and int genes are capable of forming plaques. |

4.7 Specialised transduction

Low frequency transduction

specialised
transducing
particles

If a *gal⁺ E. coli* lysogenic for λ is induced, the progeny phage may be harvested and used to infect a non lysogenic *gal⁻ E. coli*. By plating the infected cells on a minimal medium containing galactose as carbon source rather than glucose it is possible to select *gal⁺* transductants. About one transductant will be obtained for every 10^6 infecting phages.

general
recombination

Two types of transductants are produced. About one third will be stable *gal⁺* transductants arising due to general recombination between the *gal⁺* and *gal⁻* genes such that the *gal⁺* gene replaces the *gal⁻* in the recipient chromosome (see Figure 4.9).

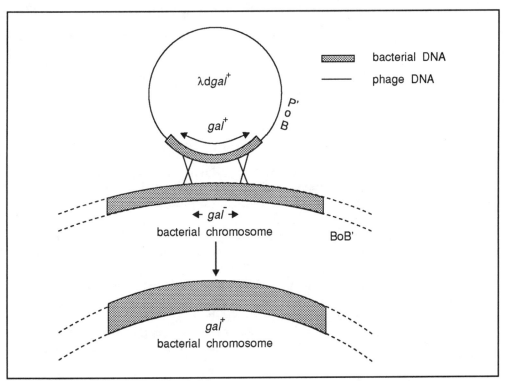

Figure 4.9 General recombination between the two *gal* genes allows the donor *gal⁺* gene to replace the recipient *gal⁻* gene.

The remaining transductants are unstable and frequently give rise to *gal⁻* cells. They are partial diploids and carry both the *gal⁺* and *gal⁻* genes and phage DNA. The reason they are unstable is that the λ d*gal⁺* phage is occasionally excised from the recipient via recombination giving rise to *gal⁻* cells. We will now describe the mechanism for this.

As λd*gal⁺* transducing particles are rare (remember that the lysate contains both λ⁺ and λ d*gal⁺*, the latter at a much lower concentration than the former) it is usual to infect *E. coli* with a high ratio of phage to bacteria. Upon doing so every *gal⁺* transductant will be simultaneously infected with a λd*gal⁺* and a wild type λ (λ⁺) . It is thought that the wild type λ first integrates at *att*B, followed by integration of λd*gal*⁺ at one of the hybrid sites (more often BoP′ than PoB′). These partial diploids are very unstable due to general recombination between the two homologous λ sequences. A cross-over event leaves a *gal*⁻ chromosome carrying a recombinant, but intact λ⁺ prophage (see SAQ 4.6).

In a similar manner, a cross-over between the two bacterial *gal* sequences will generate a *gal*⁺ chromosome carrying a λ⁺ prophage (see SAQ 4.6).

SAQ 4.6

The result of a double infection giving rise to the λ⁺ / λd*gal*⁺ transductant as described previously is drawn below.

Sketch the crossover event a) between homologous λ sequences that gives rise to λ⁺/*gal*⁻ recombinants and b) between homologous *gal* sequences that gives rise to λ⁺/*gal*⁺ recombinants.

High frequency transduction

When a λ⁺/λd*gal*⁺ lysogen is induced equal numbers of λ⁺ and λd*gal*⁺ phages are released. Both types of phage are equally viable as the defects in the λd*gal*⁺ phage are complemented by the genes on the λ⁺ prophage. For example the head and tail products needed in order to build an intact phage are provided by λ⁺. This lysate can now be used to infect a *gal*⁻ *E. coli* at a very low multiplicity of infection so that double infections are very rare. As half of the phages are λd*gal*⁺ there is a very high frequency of transduction. The λd*gal*⁺ genome inserts via general recombination between the phage *gal*⁺ and the *gal*⁻ gene of the *E. coli*, rather than via site specific recombination as the λd*gal* hybrid BoP′ site only recombines inefficiently with the normal bacterial *att* site BoB′. This contrasts with conditions after low frequency transduction, which we discussed previously, as the λd*gal*⁺ may integrate easily at the hybrid BoP′ site created following prior integration of λ⁺ upon double infection following treatment with a high ratio of phage to bacteria.

4.8 The uses of specialised transduction

As only relatively small amounts of DNA are transferred, specialised transduction is of particular value to induce small scale changes to bacterial strains such as the conversion of a *bio*⁻ strain of *E. coli* to *bio*⁺ and allows fine scale mapping or genetic manipulation. This technique is applicable to other genes of interest. Other phages have their own specific attachment sites. The phage φ80 has an attachment site next to the *trp* operon and induction of a φ80 lysogen will generate φ80 transducing particles which may be used in transduction experiments.

4.9 Generalised transduction

In contrast to the temperate phages we have just discussed, virulent phages always multiply and lyse the host cell after infection.

Generalised transduction is mediated by some virulent bacteriophages and by certain temperate bacteriophages whose chromosomes do not integrate at a specific site of attachment on the host chromosome, for example phage P1. In the latter cases generalised transducing particles are only produced during the lytic stage. The phage P1 is unusual in that following infection like λ its chromosome circularises and is repressed, but unlike λ it remains as a free supercoiled DNA molecule with one or two copies per cell. Its replication is coupled to chromosomal replication such that when the bacterium divides each cell receives a P1 prophage, only rarely (about 1 in 1000) cells in fact fail to receive a P1 phage.

generalised
transducing
particles

complete
transduction

In generalised transduction the phage mistakenly packages a piece of the bacterial host chromosomal DNA. This may happen at a frequency of 1×10^{-3} - 1×10^{-6}. Generalised transducing particles therefore contain only bacterial DNA and this DNA may be derived from any part of the bacterial chromosome. This fragment of bacterial DNA may enter a new cell and become integrated into the host chromosome following pairing and recombination with its homologous region. In this state the integrated DNA will be inherited and transduction is known as complete.

Not all virulent phage are able to mediate generalised transduction. T even phages (T2, T4 etc) for example degrade host DNA, using the nucleotides to synthesise their own DNA. Some phages do not degrade the host DNA at all so it is too big to package, others have a maturation process which is highly specific for phage DNA.

4.10 Abortive transduction

Following transduction the incoming bacterial DNA is not always recombined into the recipient bacterial chromosome. It may instead remain free and not integrated and then will not be replicated. This will be the case for example if the recipient cell is rec^{-}. Hence it will remain in only one of the two daughter cells, leading to only one cell of a colony harbouring the fragment. This is known as abortive transduction.

abortive
transduction

Identification of transduced cells employs selection techniques similar to those discussed in Chapter 3.

4.11 Use of transduction for genetic maps

Generalised transduction is often used for mapping studies. Only about 1-2% (25-30 genes) of the host chromosome is picked up by the phage and an even smaller amount retained within the recipient bacterial chromosome following recombination. Only genes that are close together on the host chromosome will undergo co-transduction. If the transduction frequency is 10^{-5}, we would expect a single cell to be transduced twice at a frequency of 10^{-10}, ie very rarely.

co-transduction

If we wanted to transduce a *his⁻ E. coli* strain to *his⁺* we could isolate P1 phage following their infection of *his⁺* bacteria and use these phages to infect *his⁻* bacteria.

When performing P1 phage transductions only one particle in 10^5 carries bacterial DNA. This means that a large number of active phages is present. It is therefore desirable to prevent these phages from infecting and lysing or lysogenising the transductants.

Given the fact that bacteriophage P1 requires Ca^{2+} to attach to *E. coli* and that Ca^{2+} can be removed from the medium by adding sodium citrate which binds it, which of the following would permit maximum recovery of transduced *E. coli*. Ring the appropriate response.

1) Phage mixed with excess bacteria. Before bacteria lyse to produce new phage remove all Ca^{2+}. Plate on selective medium lacking histidine but containing sodium citrate.

2) Mix excess phage with a small number of bacteria. Before bacteria lyse to produce new phage remove all Ca^{2+}. Plate on selective medium lacking histidine but containing sodium citrate.

3) Phage mixed with excess bacteria. Plate on selective medium lacking histidine but containing Ca^{2+}.

If two markers are to be consistently co-transduced they must be within the same fragment of donor DNA and therefore be closely linked. In contrast if they cannot be co-transduced they are probably at least 100kb apart (the approximate length of a fragment).

Consider three genes *X*, *Y*, and *Z* 70 kb apart.

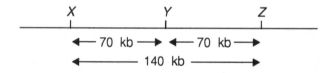

It is possible to co-transduce *X* with *Y* and *Y* with *Z*, but not *X* with *Z*.

We can transduce an *X⁻Y⁻ E. coli* with a phage carrying *X⁺Y⁺* DNA and select *X⁺* transductants on a selective medium which allows the growth of *X⁺* but not *X⁻ E. coli*. Assuming our selection process does not effect the *Y* locus we will expect either *X⁺Y⁺* or *X⁺Y⁻* transductants. The proportions we obtain can be determined by testing the ability of different transductants to grow on a medium which will distinguish the *Y⁺* and *Y⁻* phenotypes. In order to obtain an *X⁺* transductant there must be an even number of cross-overs (see Chapter 3). Let us take the simplest case involving just two cross-overs. One before *X* and the second either between *X* and *Y*, to produce *X⁺* and *Y⁻* transductants (see Figure 4.10a) or after *Y* to produce *X⁺Y⁻* transductants (see Figure 4.10b).

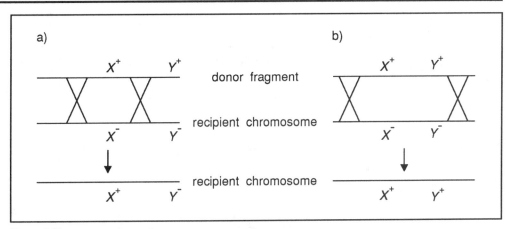

Figure 4.10 a) production of $X^+ Y^-$ transductants b) $X^+ Y^-$ transductants.

The frequency with which cross-overs are obtained between two points is in general proportional to the genetic distance between them, so it follows that the closer that genes X and Y are to each other the fewer cross-overs will occur between the, hence the smaller the number of $X^+ Y^-$ transductants that will be observed. Two genes that are abutting each other are unlikely to have a cross-over between them. This information may be used to order genes and is described in the following section on three point crosses.

4.12 Ordering genes by three point crosses

multiple cross-overs

Using data obtained from transduction studies with three selectable marker genes, it is possible to determine gene order. With three point crosses we seek to account for data with the minimum number of cross overs. The largest class of data obtained will be from double cross-overs, the smallest from multiple cross-overs (or in other words, double cross-overs are taken to be a more plausible explanation for experimental data than multiple cross-overs).

Consider genes P, Q and R. At present their order is unknown. The possibilities are:

Q between P and R	$P\ Q\ R$
R between Q and P	$Q\ R\ P$
P between Q and R	$Q\ P\ R$

Suppose we have a recipient *E. coli* of genotype P^-, Q^-, R^- which is infected with a transducing phage carrying genes P^+, Q^+, R^+. We can study the transductants by selecting one marker and looking at the nonselected markers. For example select for P^+ and look at Q and R, or select Q^+ and look at P and R.

Let us assume the order is in fact $P\ Q\ R$. If the selected marker is one of the outer markers (P^+ or R^+) the P^+, Q^-, R^+. recombinants can only be accounted for if there are four cross-over events, see Figure 4.11a.

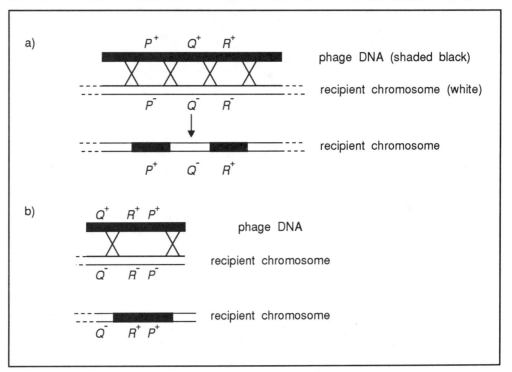

Figure 4.11 a) Four cross-over recombinational events, b) double cross-over event.

We would expect such an event to be rare.

If instead the gene order was $Q R P$ or $Q P R$ then $P^+ Q^- R^+$ recombinants would arise due to a double cross-over event, see Figure 4.11b.

Another way of putting it is, if the central marker is selected then the outside two markers will appear independently in the transductants, whereas if one of the outer two markers is selected it will be more usual for the transductants which inherit the other outside marker to also inherit the central marker.

We can determine whether double or quadruple cross-overs are likely to have arisen by looking at data from such an experiment (see SAQ 4.8).

SAQ 4.8

The following experiment was carried out:

$$E.\ coli\ 1^-2^-3^-\ \times\ phage\ 1^+2^+3^+$$

The number of transductants in each class was noted following selection for either 1^+ or 3^+.

Selecting for 1^+
number of transductants

Selecting for 3^+
number of transductants

a)	$\underline{1^+\ 2^+\ 3^+}$	91	e)	$\underline{1^+}\ 2^+\ \underline{3^+}$	69
b)	$\underline{1^+\ 2^+}\ 3^-$	105	f)	$\underline{1^+}\ 2^-\ \underline{3^+}$	230
c)	$\underline{1^+}\ 2^-\ \underline{3^+}$	236	g)	$1^-\ \underline{2^+\ 3^+}$	1
d)	$\underline{1^+}\ 2^-\ 3^-$	108	h)	$1^-\ 2^-\ \underline{3^+}$	228

The markers inherited from the donor are underlined.

1) Which category of transductants (a-h) do you think have been formed by a quadruple cross-over?

2) What do you think the gene order is?

   a) 1,2,3; b) 2,3,1; c) 2,1,3.

4.13 Transposable elements

In our study so far we have assumed that in bacterial chromosomes, plasmids and phages, genes occupy relatively fixed positions and do not move sites. However, some DNA sequences can and do change position. These mobile DNA sequences are called transposons transposable genetic elements or transposons. True transposable elements are unable to exist autonomously and can only replicate when they are a component part of an independent replicon. Such sequences are found in a diverse range of organisms such as bacteria, fungi, insects, plants and mammals. We will limit ourselves to consideration of bacterial transposable elements.

In the 1970s evidence for the existence of bacterial transposable elements was uncovered when it was discovered that the gene for ampicillin resistance was being transferred from one bacterial plasmid under study to another with a concomitant increase in size of 4.5Kb of DNA for the latter one. The ampicillin gene was contained on a mobile genetic element called a transposable element.

Transposable elements can be divided into two categories:

• insertion sequences;

• transposons;

that differ in their size, complexity and structure.

4.14 Insertion sequences

insertion
sequences

Insertion sequences (known as IS) are simple transposable elements. They are characterised by their ability to transpose from one genome site to another and by the fact that they possess only the genes involved in their own transposition. *E. coli* possesses at least 5 such elements predictably named IS1, IS2, IS3, IS4 and IS5.

inverted
repeats

All known IS elements are structurally similar. They comprise a defined double-stranded DNA sequence which is 0.7 -1.5 kb in length, at the two termini of which are a pair of nearly perfect inverted terminal repeat sequences. These inverted repeats (IRs) vary in length but are in the region of 9 - 40 basepairs long. They are thought to be the regions recognised by the proteins which promote transposition and are, as such, essential. Between the inverted repeats is found the sequence encoding the transposase, the protein which is responsible for transpositional recombination. There is often, as in IS1, an additional, smaller, separate sequence, encoding a second protein required for transposition.

transposase

target
sequences

In addition each IS is flanked by a short 3 - 13 bp, direct repeat of host DNA. It serves to identify the target sequence (TS) on the host DNA and during the transposition event it is duplicated.

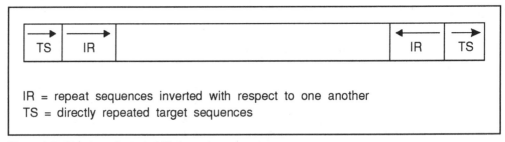

Figure 4.12 Structure of a typical IS element.

These IS elements move (transpose) once every 10^5 - 10^9 generations. Although they possess no genes other than those involved in their own transposition they do include transcriptional and translational start and stop codons and so may have a dramatic effect on the expression of more distant genes in the genome into which they are inserted. IS elements may cause a variety of chromosomal aberrations such as deletion or inversion of DNA segments immediately adjacent to the IS element.

IS elements are present on many pieces of DNA. The example we have seen in Chapter 3 involved the Hfr. Here IS elements on both the bacterial chromosome and the F plasmid provide regions of homologous DNA which allow recombinational insertion of the F plasmid into the bacterial chromosome in the formation of the Hfr. The different Hfr strains that occur are due to such F factor insertions at different sites and in different orientations as different bacterial strains harbour IS elements in variable positions.

4.15 Transposons

transposons

Transposons are distinguished from IS elements by their greater size (often 2-20 kilobases) and by the fact that they possess at least one gene, not required for the transposition event itself, such as resistance to an antibiotic. There are two classes of transposons, composite transposon or complex transposon.

4.16 Composite transposons

composite
transposons

Composite transposons are relatively simple in structure and are about 2 - 10 kb in size. They have a central region containing genes for example encoding antibiotic resistance, flanked by a pair of identical IS elements, or a pair of IS-like elements. These elements may either be direct or inverted repeats of each other. Such composite transposons can transpose themselves by the action of the flanking elements. Examples of such elements are Tn5 and Tn9.

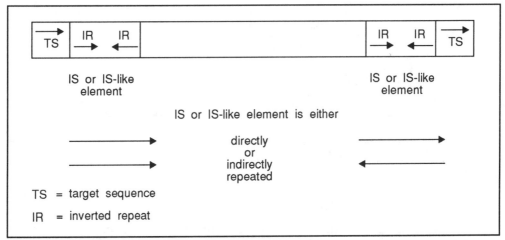

Figure 4.13 Structure of a generalised composite transposon.

Bacterial transposons are capable of carrying many antibiotic resistances around, for example:

- Tn 9 carries chloramphenicol resistance;

- Tn 5 carries kanamycin resistance;

- Tn10 carries tetracycline resistance.

At least one of the flanking IS or IS-like elements encodes the transposase protein(s) which acts on the terminal inverted repeats to promote transposition.

4.17 Complex transposons

complex
transposons

Complex transposons differ from composite transposons in that they are not flanked by IS or IS-like elements, but instead by a pair of short inverted repeat sequences, often 35-40 bp long. These IR sequences have no independent transpositional activity of their own. Between the IR sequences are two genes encoding proteins which are essential for transposition and one or more resistance genes. When complex transposons transpose they too duplicate a sequence of target DNA. A commonly studied example of a complex transposon is Tn3 (see Figure 4.14). Tn3 encodes resistance to ampicillin together with transposase and resolvase proteins, both of which are essential for transposition.

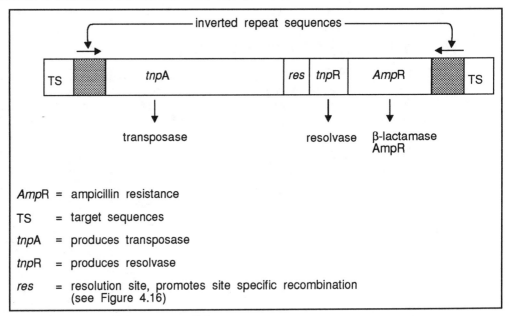

Figure 4.14 Structure of the transposon Tn3.

4.18 Transposition

replicative and
non-replicative
(conservative)
transposition

The transposition event is relatively rare (10^{-5} - 10^{-9} per element per generation for IS elements and 10^{-3} - 10^{-5} per element per generation for transposons) and enables a transposable element to move from one position in the host genome to another. Some transposable elements regularly transpose via a replicative process in which the original element stays in the same position in the genome and a new element appears at a different location. Other types of transposable element cut themselves out from their original location and insert into a target sequence on a different replicon (see Figure 4.15).

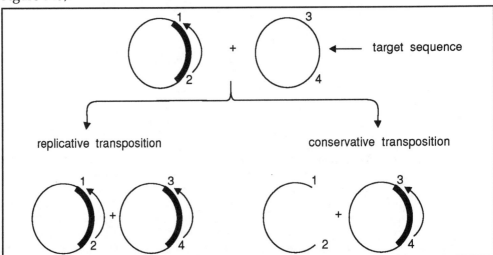

Figure 4.15 Events involved in replicative and non-replicative transposition. Regions 1-4 act as reference points.

All transposition events absolutely require both the element encoding transposase proteins and the terminal inverted repeat or internal resolution sites at which these proteins act.

The *E. coli* recombination system is not absolutely required and transposition may occur in *rec*A⁻ hosts.

There are several theories to explain the transpositional events in molecular terms, many of which are too complex for our present considerations. The transposon Tn3 is probably the best understood transposon. In simple terms its transposition can be summarised as follows. The Tn3 transposase encoded by gene *tnp*A⁺ is involved in transposition of the transposon to a new site and in the formation of a cointegrate molecule via replicon fusion (see Figure 4.16a). The *tn*R⁺ gene product, resolvase, binds to sites called *res* sites on the two copies of Tn3 which are aligned as in Figure 4.16b and site specific recombination occurs via reciprocal cross-over in the *res* sites. The two original replicons are regenerated but now each has a copy of Tn3.

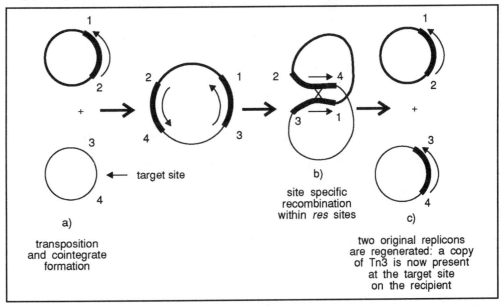

Figure 4.16 Transposition of Tn3. Note thick bars represent the transposon. (See text for a description).

SAQ 4.9

Indicate which of the following statements are true or false:

1) composite transposons are flanked by IS or IS like elements;

2) complex transposons are flanked by short inverted repeat sequences;

3) composite transposons contain only two inverted repeat sequences;

4) Tn3 is a complex transposon encoding kanamycin resistance;

5) Tn9 is a composite transposon encoding chloramphenicol resistance;

6) the *tnp*R gene of Tn3 encodes a resolvase;

7) the *tnp*A gene of Tn3 encodes a β-lactamase for ampicillin resistance.

Summary and objectives

In this chapter we have examined transduction and transposition. In order to understand transduction, we described the two life cycles found amongst bacteriophages and explained how specialised and generalised transducing particles may be produced. We described how the analysis of the products of transduction enable us to map genes especially those in close physical proximity to each other. We also described genetic elements (transposons and insertion sequences) which can migrate around the genome.

Now that you have completed this chapter you should be able to:

• draw a diagram of a simple bacteriophage such as λ;

• understand what is meant by lysogeny;

• describe what happens during the λ lytic cycle;

• be aware of the different λ gene orders in the phage head, in the circular molecule formed upon infection and in the prophage during lysogeny;

• be aware of the importance of excisionase and integrase in the lysogenic life cycle;

• define transduction;

• contrast specialised and generalised transduction;

• be able to sketch how λd*bio* and λd*gal* specialised transducing particles are formed;

• contrast high and low frequency transduction;

• describe abortive transduction;

• be aware of the relationship between the cross-over frequency and the distance apart of two markers;

• understand how to map genes by three point crosses;

• describe and sketch a typical IS element, composite transposon and complex transposon;

• draw a figure to illustrate the mechanism of Tn3 transposition.

Gene structure and expression: transcription in prokaryotes

Gene structure and expression: transcription in prokaryotes

5.1 Introduction

5.1.1 Historical perspective

All cells have DNA as their genetic material. The information contained in DNA will ultimately code for and control all the components of the cell. The double helix model proposed by Watson and Crick in 1953, elegantly suggested how DNA could be both an information store and undergo accurate replication for transmission of that information. What was the basis of these conclusions? Briefly, the complementary pattern of basepairing (A with T, G with C) suggested that each strand could act as the template to make the alternative DNA strand. If the sequence of bases on an individual strand was read in triplets, then there were more than enough codons to represent each known amino acid (see Chapter 1).

By 1957, Ingram had shown that the inherited human disease sickle cell anaemia resulted from a defective haemoglobin molecule. A single amino acid change, (from glutamic acid to valine), in one of the haemoglobin polypeptides, was the cause of this severe genetic disease. So, as this disease was inherited, presumably a change must have occurred in the DNA which was responsible for this amino acid change. However, in humans, the DNA is physically separated from the ribosomes in the cytosol (the site of protein synthesis) by the nuclear membrane and hence an intermediate molecule had to be involved in translating the information from DNA into the eventual protein that was coded by the defective gene.

Why would bacteria have not initially provided evidence of this kind?

In prokaryotes, the locations of DNA (genetic information) and protein synthesis, (the product) are not separated by a nuclear membrane. Hence, it would have been less compelling that there had to be an intermediate carrier molecule.

What was the intermediate information carrier from DNA to protein? Suspicion had immediately fallen on RNA. The close resemblance of RNA and DNA provided a clear possibility of a single-stranded DNA acting as a template for the synthesis of a complementary RNA molecule.

5.1.2 The central dogma

central dogma   By 1956 Crick had formulated these ideas into one of the main planks of modern biology. The so-called 'central dogma' proposed a one way flow of genetic information from DNA, through RNA to protein:

So, there were two essential processes which converted the information in DNA into protein. Transcription produced an RNA copy which was then translated into protein.

Whilst this theory remains essentially unchallenged today, we do know of some rare exceptions to the rule of one way flow of information from DNA to RNA. We shall briefly consider the nature of these exceptions towards the end of the chapter.

The early molecular biologists turned to the simple microbial systems of viruses and bacteria to explore the process of transcription. These systems were already well developed in providing a wealth of genetical and physiological model systems.

The rest of this chapter will consider the basic principles of transcription in prokaryotic systems. Although many features of eukaryotic transcription differ from that in prokaryotes the underlying mechanism of transcription is closely related in both cell types. We can define transcription as the synthesis of a single-stranded RNA molecule from a double-stranded DNA template, catalysed by a specific enzyme; RNA polymerase. First we shall briefly think about some of the features of RNA that distinguish it from DNA.

5.2 RNA

5.2.1 General characteristics

Two main differences exist between the chemical structures of typical DNA and RNA molecules:

- the sugar residue of DNA is deoxyribose whilst in RNA the presence of an extra hydroxyl group (-OH) at the 2′ position of the carbon ring, produces the sugar ribose;

- the pyrimidine base uracil (RNA) replaces thymine (DNA).

Look at Figure 5.1 which illustrates these changes. These relatively small changes mean that RNA can still form complementary double helices with DNA to form DNA/RNA hybrids. However, RNA is most commonly found as single-stranded molecules.

∏ What are the key differences in the structure of thymine and uracil? How does deoxyribose differ from ribose?

Thymine is a slightly heavier molecule than uracil - a methyl (CH_3) group is substituted for a hydrogen atom (H). Deoxyribose does not have the additional oxygen atom at (carbon) position 2′. You may want to identify these differences by drawing arrows on the figure.

The 2′ -OH group of ribose is left free when ribonucleotides link to form RNA, so making the molecule inherently less stable than DNA. In aqueous solutions, RNA undergoes hydrolytic cleavage much faster than DNA. All of this makes RNA a much more difficult molecule to study than DNA in the laboratory.

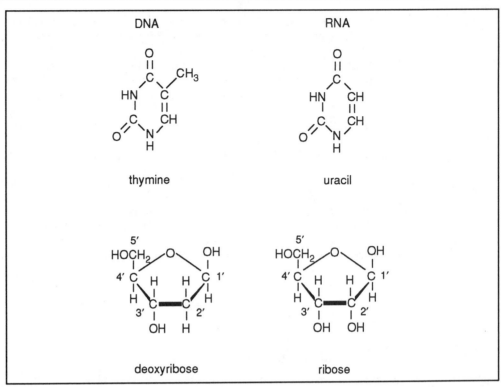

Figure 5.1 Differences in the chemical structures of DNA and RNA. Thymine differs from uracil in one position. Deoxyribose differs from ribose in one position.

5.2.2 Messenger RNA

Before Watson and Crick proposed their double helix model for DNA, Chargaff established that the ratio of bases A to T or G to C was always close to one. The double helix model proposed that A was paired or bonded with T and G with C, so preserving this strict ratio. Similarly, it appeared reasonable to suggest that if DNA transferred information to RNA, then a DNA/RNA hybrid intermediate may exist with A pairing with U and G with C.

molecule/ base	Uninfected E. coli DNA	rRNA	T2 phage DNA	RNA source?
A	25	25	32	31
U/T	25	21	32	29
G	25	32	18	18
C	25	22	18	22

Table 5.1 Base composition of DNA and RNA of E.coli and T2 (%). RNA source? is the RNA produced immediately after infection of E.coli with T2 phage. See SAQ 5.1.

When the phage T2 infects *E.coli*, bacterial RNA and protein synthesis soon stops and phage protein synthesis begins. Immediately after phage infection, there is a rapid synthesis followed by degradation of a small amount of RNA. Look at Table 5.1 which shows the results of an infection experiment in which Volkin and Astrachan measured the base compositions of different nucleic acids.

RNA source?, represents the data for the RNA that was produced immediately after infection of *E. coli* with T2. Which DNA molecule is responsible for this new RNA and what is the significance of this observation?

On the basis of such experiments, the phrase 'messenger RNA' (mRNA) was used to describe this type of RNA that was rapidly turned over in the cell and carried information from DNA to protein. So, we can now clearly differentiate mRNA, involved in transcription, from the two other key types of RNA involved in translation:

- rRNA, ribosomal RNA forming a structural component of ribosomes;

- tRNA, transfer RNA which brings individual amino acids to the growing polypeptide chain.

5.2.3 RNA polymerase

bacterial RNA
polymerase

holo-enzyme

More detailed experiments of the type outlined in SAQ 5.1 became possible with the discovery of an enzyme system capable of assembling RNA molecules on a DNA template. In each case the base composition of mRNA molecules closely resembled the DNA of the organism. The enzyme was called DNA-dependent RNA polymerase and required ribonucleoside triphosphates (ATP, GTP, CTP and UTP) as precursors. In bacteria, this same enzyme is responsible for making messenger, transfer and ribosomal RNA from specific DNA sequences. Bacterial RNA polymerase is made from a number of subunits. The active form (holoenzyme) in *E. coli* has five different polypeptide chains. These are given the following symbols and have the molecular weights as specified below:

- β' 155 000 Daltons;

- β 151 000 Daltons;

- σ 70 000 Daltons;

- α 36 500 Daltons;

- ω 11000 Daltons.

Each chain is represented once per holoenzyme, apart from α which has two copies per holoenzyme. This makes a very large molecule indeed. What might the size of this enzyme convey to us? If this enzyme is making an RNA copy of a DNA template then presumably the polymerase can physically cover a significant length of DNA sequence. In fact, the holoenzyme can bind to over 60 nucleotides at a time. We might like to suggest that the large size of the enzyme also increases the surface area available for interactions with other molecules in addition to DNA. Interestingly, phage T7 RNA polymerase consists of only a single chain. We are not yet sure of the precise functions of all of the *E. coli* enzyme subunits. However, the antibiotic rifampicin which is a potent inhibitor of bacterial RNA polymerase, binds to the β subunit and prevents RNA chain initiation but not elongation. This, together with other evidence, indicates that the β subunit holds the key catalytic functions of the holoenzyme.

We know that in some bacteria, different sigma (σ) factors exist. For example, in *Bacillus subtilis* there are at least 4 different types. Binding of the different sigma factors allows expression of different genes by recognition of different DNA sequences.

SAQ 5.2

When an *E. coli* cell is growing at 37°C it has been estimated that at any one time, the chromosome is transcribing:

up to 800 unique mRNA molecules;

100 different tRNA molecules;

700 rRNA precursor molecules.

1) If a single transcript requires a single RNA polymerase molecule then how many RNA polymerase molecules are at work?

2) If only 4% of the total cell RNA is mRNA, why does it appear to make up 50% of the transcribed RNA?

SAQ 5.3

1) If the average rate of transcription in *E. coli* is 60 nucleotides per second, how long would it take to produce an mRNA of 2000 nucleotides in length?

2) A specific nucleotide remains incorporated within the mRNA molecule for only about 90 seconds. Comment on the significance of this observation for translation of the mRNA.

5.3 Pattern of transcription

5.3.1 Transcription of one strand only

Although the DNA molecule consists of a double helix, only one of these strands acts as a coding strand for a given RNA molecule. This was first shown using the phage SP8 which infects the bacterium *Bacillus subtilis*. Each strand of the SP8 DNA can be distinguished on the basis of the 'heavy' purines being concentrated on one strand. Adenine and guanine have higher molecular weights than the pyrimidines cytosine and thymine. The two strands can be physically separated by density gradient centrifugation.

Look at Figure 5.2 which shows the basic idea. The method relies on the system producing DNA/RNA hybrid molecules. Hybridisation is the process by which two single-stranded molecules come together to form hydrogen bonds between segments with complementary nucleotide sequences. In the case illustrated in Figure 5.2, the mRNA produced after SP8 infection hybridises with the heavy SP8-DNA strand. Only the heavy strand of SP8 DNA could have been the template for the SP8 mRNA forming a hybrid of one chain of DNA linked to one chain of RNA.

coding
strand(s)

It is important to remember that on a long molecule of DNA like an *E. coli* chromosome, one strand will serve as the coding strand for some genes, whilst the other strand will be the coding strand for the other genes.

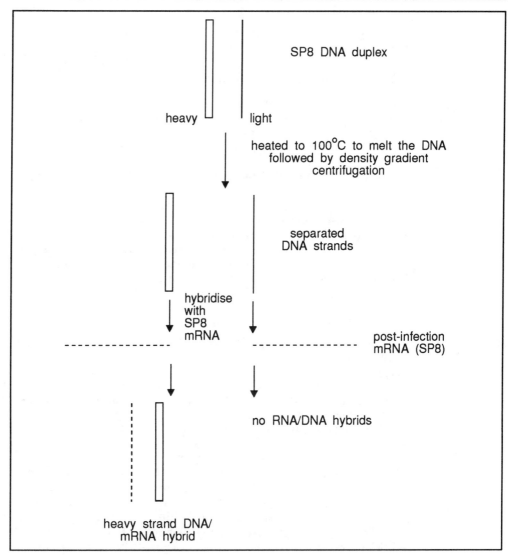

Figure 5.2 An experiment using *Bacillus subtilis* phage SP8, to show that only one DNA strand acts as a template for mRNA.

5.3.2 Synthesis of RNA in the 5' → 3' direction

RNA is
synthesised
5' → 3'

We can identify each end of an RNA molecule in much the same way as a DNA molecule. One end is terminated by a 5' carbon atom (relative to the ribose ring), whilst the other end finishes with a 3' carbon atom. The RNA molecule is always synthesised on the DNA template in a 5' → 3' direction. This was first shown by growing *E. coli* cells at 0°C in the presence of a ^{14}C labelled uracil precursor, uridine. At this low temperature, RNA synthesis is slowed to the rate of one nucleotide addition per 13 seconds. When the growing mRNA was extracted it was shown that the ^{14}C label appeared at the 3' end of the growing RNA chain. So the molecule must be growing in the 5' → 3' direction.

So, having established that only one strand of DNA is transcribed for a given sequence and RNA synthesis proceeds in a 5' → 3' direction, how does the RNA polymerase recognise the site to start transcription, ie the beginning of a gene?

5.4 Determination of RNA polymerase binding sites

5.4.1 Polymerase recognition functions

The recognition of specific sites on the DNA molecule to begin transcription basically depends upon two factors:

- identification by the sigma subunit (σ) of RNA polymerase;

- possession by the DNA binding site of a specific sequence; the promoter.

The sigma subunit can easily be detached from the holoenzyme to yield the core enzyme. By itself, the core enzyme can polymerize RNA using a DNA template, but the reaction is inefficient and transcription starts at seemingly random points along the DNA molecule. Although the sigma subunit has no catalytic activity on its own, if it is added to the core enzyme, the reconstituted holoenzyme can then start RNA synthesis at specific sites. These specific sites represent the beginnings of distinct genes or operons (groups of clustered genes with a common promoter).

5.4.2 DNA sequences of promoters

promoter

footprinting

It is likely that the nucleotide sequences of a promoter give the DNA a specific, recognisable shape, into which the RNA polymerase can fit. Promoters can be recognised by using the technique of footprinting. Fragments of chromosomal DNA derived from the 5' end of a gene are mixed with purified RNA polymerase *in vitro*. As this is a controlled, artificial test tube procedure transcription does not occur but the RNA polymerase can still bind to any promoter sequences. The mixture is then treated with DNase which will digest all the DNA, except that protected by binding to RNA polymerase. These 'footprints' can then be characterised by sequencing the protected DNA. Although conceptually simple, footprinting can nevertheless be a lengthy and difficult procedure and recombinant DNA technology offers further methods of isolating promoters. These centre on fusing test DNA sequences which may have promoters to reporter genes ie genes which have no promoters of their own. A promoter from the test DNA can 'switch on' the reporter gene by initiating transcription. Transcription of the reporter gene can be easily observed, for example if it codes for production of a pigment, or confers antibiotic resistance on the cell.

detecting promoters in a reporter gene

consensus sequences

When promoters are examined in detail, it is found that they contain common elements, involved in RNA polymerase binding, called consensus sequences. These short, conserved sequences are essential to promoter function. This can be confirmed when mutant promoters are sequenced. Loss of promoter function results from changes in the consensus sequences. The consensus sequences are found just before (upstream) of the actual initiating nucleotide for the transcribing RNA molecule. In *E. coli* two sequences are particularly important:

- the region - 35 basepairs from the transcription start with a sequence of TTGACA;

- the region at -10 basepairs from the transcription start with a sequence of TATATT.

5.4.3 Promoter strength

strong and
weak
promoters

Interestingly, promoters that are relatively weak in terms of showing a low level of transcription of their genes often have sequences that differ markedly from the consensus sequences. Strong promoters tend to match the consensus sequences more closely. Thus, promoters can be defined as weak or strong depending on the frequency with which the RNA polymerase binds to successfully produce an RNA transcript. The level of transcription of a gene is closely correlated with the strength of binding between the RNA polymerase enzyme and the promoter.

Computer simulations can help us visualise the complex process of RNA polymerase binding to the DNA molecule in a three dimensional way. The commonest form of DNA (B-form) has around 10 basepairs per turn of the helix. As the -35 and -10 'boxes' are separated by about 20 basepairs, this represents two complete turns of the DNA helix and so the RNA polymerase must bind along one face of the DNA molecule. Other sequences further upstream of the -35 box along with other proteins can also influence the success of binding. We shall consider these factors in a later chapter.

SAQ 5.4

Look at Figure 5.3 which shows the base sequence of 3 different promoters relative to the transcription start site. Identify the consensus sequences in promoter A and draw a box around them. Use promoter a now to identify and box the consensus sequences in promoters b and c. Are they identical? Remember, deviations from the consensus sequence may affect promoter strength.

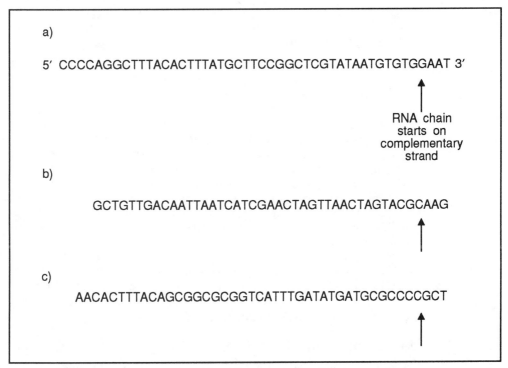

Figure 5.3 Sequences of promoters from three different genes. Enclose the two consensus sequences of each promoter in a box (see SAQ 5.4).

Remember, that RNA is synthesised in the 5′ → 3′ direction and so transcription begins at the 3′ end of the 'sense' strand of DNA. By convention molecular biologists show the sequence of DNA as the complementary (antisense) strand. This convention means that we can then write the RNA sequence coded by the gene as identical to the DNA sequence, except that T is exchanged for U in RNA (see Chapter 2).

5.5 Transcription *in vivo*

5.5.1 RNA polymerase binding

Having considered the main features of transcription, we shall now try to think of transcription as a dynamic process with a number of key steps.

RNA polymerase recognises a specific promoter through its sigma subunit and binds to this DNA sequence. The *E. coli* enzyme shows two steps in this binding process;

- initial recognition occurs at the -35 region resulting in loose binding;

- the RNA polymerase binds more tightly as the double helix begins to unwind in the -10 region. DNA/DNA basepairing breaks down locally.

This produces an 'open' structure which exposes the DNA template. As it is necessary for the RNA polymerase to unwind the two DNA strands, the degree to which the DNA helix is already further coiled affects the efficiency of the promoter. Enzymes which alter this 'supercoiling' (topoisomerases) can thus affect promoter activity.

5.5.2 RNA chain initiation

We can compare transcription to the flow of a river - it starts 'upstream' with RNA polymerase recognition at -35 and continues 'downstream' towards -10 and eventually to the start point of mRNA synthesis.

Once the RNA polymerase has bound tightly to the promoter and an 'open' complex has formed, then RNA synthesis - chain initiation, can begin. The first nucleotide to be added in the new RNA chain is usually A or G, with C and U being much less common.The incoming base has a triphosphate group attached. Initiation occurs around 12 to 13 bases downstream from the start of the -10 box in the promoter region but the enzyme appears to prefer to insert purine rather than a pyrimidine at the start site. It is often difficult to precisely locate the initiating base as RNA molecules can later be cleaved by RNA degrading enzymes. So the molecule that we isolate from a cell may not be the same as the original transcript.

Once the RNA chain has been started, the sigma subunit detaches from the RNA polymerase/DNA complex. It may be that sigma is required for promoter recognition elsewhere or its loss may be necessary for a weaker association between RNA polymerase and the DNA template required for the next stage.

5.5.3 RNA chain elongation

The synthesis of the RNA chain continues by the addition of ribonucleoside monophosphates. Incoming ribonucleoside triphosphates find the 3′-OH group of the end of the RNA chain and polymerise with it for the loss of two phosphate groups. As the RNA chain extends, RNA polymerase unwinds the DNA helix ahead. For a short region (approximately 12 basepairs), a DNA/RNA hybrid forms obeying basepairing rules, until the RNA strand leaves the DNA template allowing the DNA double helix to reform.

Π Figure 5.4 illustrates this process. Draw an arrow on the diagram to indicate the direction of transcription (you can check your answer in the legend).

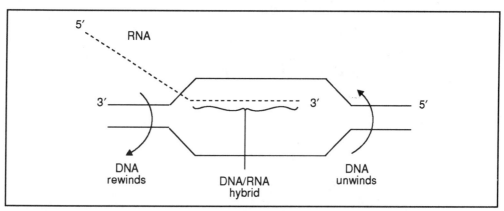

Figure 5.4 RNA chain elongation. RNA polymerase extends over the whole region drawn, with perhaps 30 nucleotides being covered by the enzyme/DNA template/mRNA complex. Transcription proceeds from left to right.

Although transcription can occur at a theoretical rate of up to 60 nucleotides per second, in practice the rate of RNA elongation is not constant. Elongation may consist of a series of 'pauses' interspersed with periods of rapid polymerisation. Likewise, the amount of DNA helix which is unwound ahead of the RNA chain varies and can depend on the specific DNA sequence in that region.

Antibiotics can provide us with some useful tools to study elongation:

- Streptolydigin binds with the β subunit of RNA polymerase and can block elongation. So the β subunit is involved with both chain initiation and elongation, (see Section 5.2.3).

- Actinomycin D blocks elongation by complexing with the DNA template and so stopping movement of the RNA polymerase.

5.5.4 RNA chain termination

How is chain elongation stopped?

Transcription of a gene is completed when:

- chain elongation ceases;

- the RNA transcript is fully released;

- the RNA polymerase enzyme is released.

There are at least two mechanisms in prokaryotes which achieve termination. These we will discuss below.

independent termination

Independent termination. The termination site is characterised by a G C rich inverted repeat sequence followed by a run of A's on the coding strand. The repeat sequence makes the RNA transcript form a 'hairpin loop' on itself which causes the RNA polymerase to pause, perhaps by prematurely lifting the second half of the repeat in the RNA molecule away from the DNA. The remaining A's on the DNA template form only weak bonds with their complementary U's on the RNA molecule. Consequently, the RNA transcript may literally 'fall off' the DNA template.

Π Figure 5.5 illustrates this termination process. Draw a box around the G C repeat sequence and the run of A's on the DNA template.

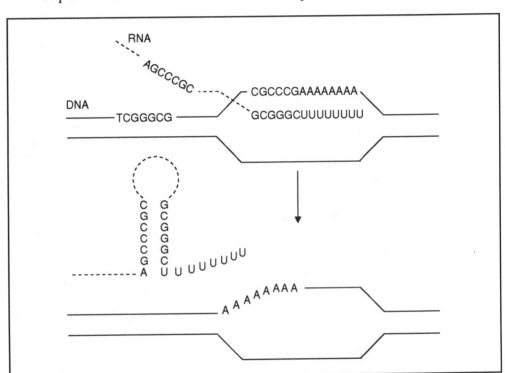

Figure 5.5 A model for independent termination. An inverted repeat sequence transcribed on the RNA strand forms a hairpin loop by complementary basepairing, destroying part of the DNA/RNA hybrid. The weak A/U bonds dissociate. Modified from Platt T, 1981 Cell 24, 10.

Once again mutations that disrupt the termination process can be shown to affect these key conserved sequences. Even changing just one of the A bases for another base can inactivate the terminator.

factor dependent termination

Factor dependent termination. Some terminators lack the run of A's and many do not seem to induce the RNA to form strong hairpin loops. In *E. coli*, if the RNA polymerase pauses, (say by formation of a weak hairpin in the RNA), then a specific protein may bind, possibly to both the RNA molecule and the enzyme. This 'rho' (ρ) protein induces the RNA transcript to fall off the DNA template and involves the hydrolysis of ATP. We do not understand this mechanism of termination particularly well - it is likely that other termination factors remain to be found.

| **SAQ 5.5** | We will now try to summarise the process of transcription *in vivo*. Arrange each of the key stages below into the correct sequence. Then use your sequence to label Figure 5.6. |

1) DNA helix unwinds as RNA chain elongates;

2) RNA chain initiates on a purine residue;

3) recognition of a promoter by RNA polymerase (-35 region);

4) RNA forms a hairpin loop and falls off the DNA template;

5) sigma subunit detaches from RNA polymerase/DNA complex;

6) DNA helix unwinds in -10 region.

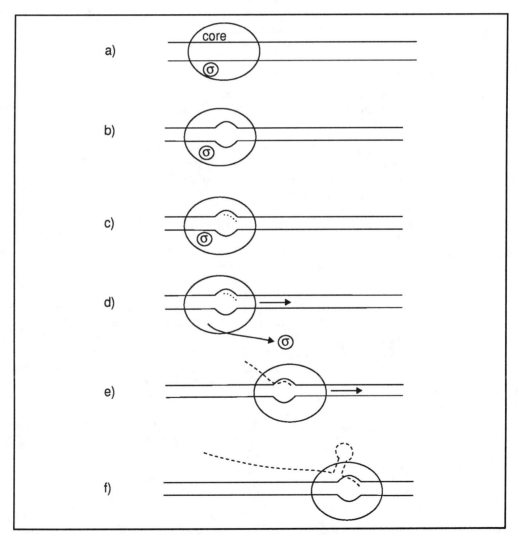

Figure 5.6 A simplified summary of stages in prokaryotic transcription.

5.6 Processing of transcription products

5.6.1 Leader sequences

The initial amino acid in a polypeptide is generally methionine. In mRNA the base sequence coding for methionine is AUG, with A being at the 5' end of the molecule transcribed in the 5' → 3' direction as we saw earlier. However, this does not necessarily mean that AUG will be found at the extreme 5' end of every mRNA. Between the actual 5' end and the initiating codon for translation (AUG) is found a series of nucleotides constituting a leader sequence. What is the function of the leader sequence? Even though its length can vary greatly between different RNAs, its common role is to establish a relationship with a ribosome for translation. In prokaryotes, a distinctive ribosome binding sequence is found. This Shine-Delgarno sequence, named after its discoverers, is situated 5-10 basepairs upstream of AUG. Not surprisingly, this sequence is homologous to the 3-OH end of 16S rRNA. It seems clear that the Shine-Delgarno sequence of the leader basepairs with the complementary base sequence of the small ribosome subunit, (3' AUUCCUCCA 5'). This then exposes the initiating codon (AUG) for correct translation of the mRNA into a polypeptide.

Shine-Delgarno sequence

5.6.2 Polycistronic mRNA

Individual genes were identified in many microbes by what is known 'cis-trans' complementation tests. So the term 'cistron' was adopted to become synonymous with the 'gene'.

operon

Many bacterial mRNAs are very long as genes may be grouped together in operational units termed operons. This mode of transcription has important implications for the control of gene expression, as we shall see in Chapter 7. Genes encoding the enzymes of a specific metabolic pathway or structural units for ribosomes can occur consecutively on the DNA molecule. A typical operon would have a single promoter which controlled a number of genes 3' to it, although examples with more than one promoter are known. The implication of the possession of a single promoter by a number of genes, is that the genes will be transcribed together on a single mRNA by RNA polymerase. This is polycistronic (ie 'many genes') mRNA - a single RNA molecule of this type may code for up to 20 proteins!

polycistronic mRNA

We have already emphasised the relative instability of prokaryotic mRNA (see SAQ 5.3), which makes it difficult experimental material. Some phages have mRNA containing complex secondary structures which increase the stability of the molecule. This makes them more attractive systems for study. Study of such systems has revealed that between the coding regions of polycistronic mRNAs are found intercistronic regions which do not appear to code for proteins and may be greater than 100 nucleotides in length in some phage RNAs. Like the leader sequences, although they are not protein coding, they help in the processing of the transcript. Normally, the polycistronic mRNA is translated as an intact molecule by ribosomes which attach to the individual translation start signals of the separate genes (cistrons).

intercistronic regions

SAQ 5.6	Look at Figure 5.7 which shows the production and translation of a polycistronic mRNA. Is it possible to predict how many polypeptides will be produced eventually from this mRNA?

Look at Figure 5.7 which shows the production and translation of a polycistronic mRNA. Is it possible to predict how many polypeptides will be produced eventually from this mRNA?

What is happening at each end of the mRNA?

When you have answered this, be careful to read our response. It contains some valuable additional information and will introduced you to the term polyribosome.

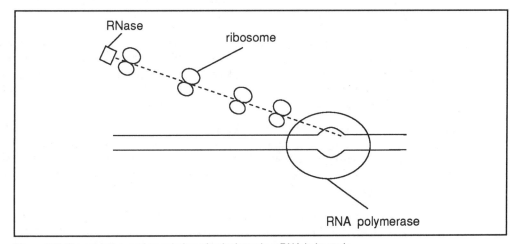

Figure 5.7 Transcription and translation of polycistronic mRNA in bacteria.

5.6.3 Processing rRNA genes in *E. coli*

The DNA sequences coding for rRNA in *E. coli* make up around 0.4% of the *E. coli* chromosome. However, rRNA accounts for 40 - 50% of RNA synthesised in growing cells. How can we explain this apparent contradiction? Clearly, the rRNA genes must be transcribed at a very high rate when compared with other structural genes. It is possible to produce 60 molecules of 30S pre-rRNA every minute from a single gene. The S value describes the rate at which the ribosomal subunit sediments during ultracentrifugation, and depends largely on its size and shape.

In bacteria, 23S and 5S rRNAs are found in a large subunit (which sediments at 50S). A 16S RNA is found in a smaller subunit (which sediments at 30S). The intact bacterial ribosome sediments at 70S. The sedimentation coefficients of subunits are not necessarily additive.

rrn operons

In *E. coli* there are seven units dispersed along the chromosome which make rRNA. These transcription units are arranged in operons producing polycistronic RNA. Each of these *rrn* operons has at least two promoters. Often it is unclear why there should be more than one promoter per gene or operon. In this case, we can suspect a numerical advantage in increasing the frequency of attachment of RNA polymerase. In other cases, complex regulatory networks may be operating to control transcription via alternative promoters. We may also find it hard to characterise potential promoters if their sequences differ markedly from the consensus sequences (see SAQ 5.4).

Each *rrn* operon contains coding sequences for:

- 16S rRNA;

- 23S rRNA;

- 5S rRNA;

- one or more tRNAs.

Each coding region is separated from the next by a transcribed spacer which will not be found in the mature RNA molecule.

The two promoters are around 120 nucleotides apart and each has normal -35 and -10 boxes, allowing two potential RNA polymerase binding sites. Transcription can initiate 7 or 8 nucleotides after the first or second promoters. The transcript appears to be terminated by the induction of a hairpin loop and a run of U's in the transcribed RNA forming a standard independent termination system.

RNA processing enzymes

The *rrn* operons are transcribed into a single RNA precursor which is then processed. This processing relies on the activity of specific RNA processing enzymes. Much of our data concerning processing comes from the study of *E. coli* mutants that are defective in these enzymes. We can then determine the role of a particular enzyme basing our conclusions on the defects we see in the processed RNA.

Figure 5.8 shows the organisation of one of these operons and the pattern of transcription and processing. In addition to having two promoters you can see that this operon has two terminators. Repeat sequences in the RNA allow the formation of the 'loop' and 'stem' structures.

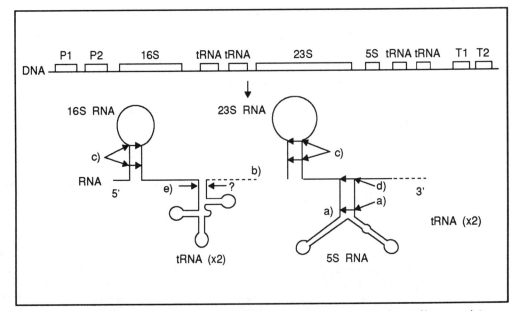

Figure 5.8 The *E. coli rrnC* operon and a simplified illustration of the processing of part of its transcript (see text for further details). Letters a) - e) refer to particular RNases described in the text. Note that P1 and P2 are promoters, T1 and T2 are terminators.

We shall briefly summarise the processing of each independent RNA from the single transcript by considering some of the enzymes involved:

- RNase III attacks a cleavage site in the spacer between the 16S and 23S sequences as the transcribed 16S and 23S RNAs form hairpin loops. Further processing to mature 16S and 23S RNA is performed by RNase M enzymes.

- RNase E generates a precursor of 5S RNA which is then further modified by another RNaseM.

- RNase P is a complex ribonucleoprotein consisting of both a polypeptide and an RNA component. This enzyme is involved in the maturation of the 5' end of tRNA although we are less clear as to the enzymes which cleave the 3' end. RNase P, then, provides us with an interesting example of how RNA itself can possess catalytic activity.

Π Now look again at Figure 5.8 and use the information provided to label the arrows with appropriate enzymes for processing the transcript.

You should label a) an RNase M, b) RNase III, c) RNase M, d) RNase E, e) RNase P.

Even this brief summary has shown that RNA processing in *E. coli* is a complex procedure which we are still trying to fully understand.

SAQ 5.7

Often the first step to understanding a complex process in a cell is to try to isolate a mutant that is defective in some state of this process. We can think of geneticists as being always one step removed from reality! If we isolate a series of mutants, each defective at a different stage, we can then build up a picture of how the normal process works. It is rather like only starting to understand how your car works when it breaks down and you have to fix it!

The following are outline phenotypes of some bacterial transcription mutants. What are the altered stages of transcription in each case?

1) rifampicin resistant;

2) streptolydigin resistant;

3) accumulates an unusually large RNA (9S) from the *rrn*C cluster.

5.6.4 Specialised forms of RNA processing

We have seen how separate functional units in polycistronic mRNAs can be delimited by intercistronic, spacer regions which do not eventually code for polypeptides (Section 5.6.2). These spacers play a role in the recognition and eventual processing of individual gene products. However, in some viruses and in Archaebacteria a more complex form of RNA processing is found. Here, an individual coding region (cistron), representing a single gene product, is itself broken by a region that is not represented in the final introns   polypeptide product. This intracistronic region is called an intron. Introns are not found in most prokaryote genomes. A pre-mRNA (or primary transcript) is processed to remove the intron before translation. As introns are typical features of eukaryotic genes they are dealt with in more detail in the BIOTOL text, 'Genome Management in

Eukaryotes'. Look at Figure 5.9 which outlines the principle of introns and intron exclusion.

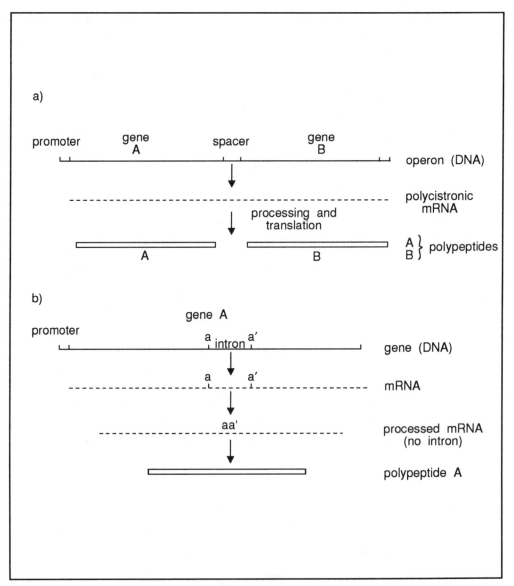

Figure 5.9 A demonstration of processing: a) Polycistronic mRNA, b) mRNA containing an intron. In a) the non-translated region is **between** genes, in b) the non-translated region is **within** genes.

5.6.5 Evolution and transcription

Archaebacteria are a very unusual group of prokaryotes with many odd features such as:

• no peptidoglycan in the cell wall;

- unusual plasma membrane lipid composition;

- eukaryote - like RNA polymerases;

- tolerance of harsh environments eg high salt or temperature.

However, despite some of these eukaryotic features, representatives of Archaebacteria are amongst the oldest known forms of life. As viruses are obligate pathogens, we could suppose that those forms which had introns could have gained them by acquiring DNA from host eukaryotes. This does not explain the presence of introns in primitive bacteria if we see introns as recently evolved features of eukaryotic cells.

Observations of this type have led us to rethink some basic assumptions about early evolution. We now believe that a simple cell (the progenote) gave rise to three separate lines or kingdoms:

- Archaebacteria;

- Eubacteria;

- Eukaryotes.

We can then suppose that introns were present in the early genetic material of the progenote. In modern bacteria (eubacteria) there has been intense selection for rapid growth and so rapid DNA replication. So, we could see how evolving eubacteria may have lost introns as excess DNA to replicate.

We saw earlier how RNA could possess enzymatic activity when we looked at RNase P activity in *E. coli* (Section 5.6.3). Introns can show self-splicing activity which is 'auto-catalytic'. This discovery has led researchers to suggest that RNA was the first 'living' self replicating molecule before cells evolved. Clearly, DNA has advantages of stability (Section 5.2.1) and so eventually this would become the commonest form of inherited information. These ideas will lead us on now to consider some modern day examples of RNA acting as the genetic material.

5.7 Modifications of the central dogma

5.7.1 RNA as a template for RNA

In Section 5.1.2 we described the central dogma of Crick, which represented a one-way flow of information from DNA to RNA and so to protein. By the late 1950s it had been shown that tobacco mosaic virus RNA could cause disease in the absence of viral proteins. A number of other viruses, like polio and influenza, have now been shown to have RNA as their genetic material. Replication of the RNA of these viruses occurs by complementary basepairing between two chains of RNA molecules.

5.7.2 Reverse transcriptase

Not all RNA viruses are limited to replication via double-stranded RNA. Retroviruses, which include RNA tumour causing viruses and HIV (the causative agent of AIDS), can synthesise a single-stranded DNA complement from their RNA template. The double-stranded version of this DNA can then be inserted into the host chromosome and produce an RNA molecule identical to the original template, upon transcription. Figure 5.10 summarises this process. The enzyme responsible for this flow of genetic information from RNA to DNA is reverse transcriptase (also called RNA dependent DNA polymerase).

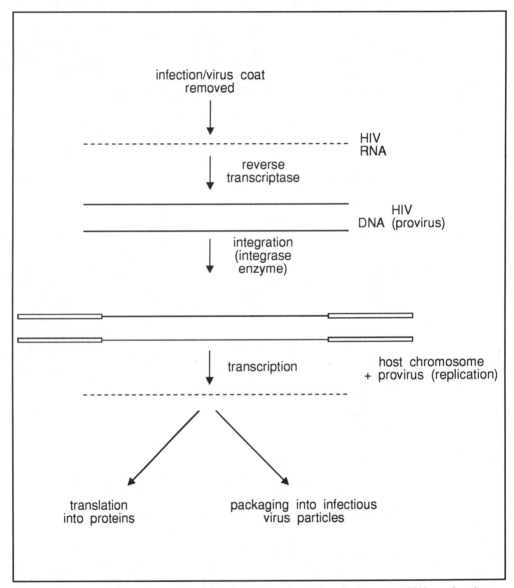

Figure 5.10 The HIV genome in its different forms. RNA produced from the integrated DNA can function as mRNA or as an RNA genome.

In the case of HIV infection, stimulation of latently infected cells, which have integrated viral genomes within the human chromosomes, results in transcription of HIV RNA. This RNA can then act as a template for viral protein synthesis or as an RNA genome for packaging into more infectious particles. The *pol* gene of the HIV virus itself codes for the reverse transcriptase. Clearly, it is a crucial research goal to try to understand and control these transcription - centred processes.

Early attempts to control the replication of HIV have tried to develop drugs which inhibit reverse transcriptase. Some success has been achieved with drugs like AZT in delaying the onset of AIDS in HIV positive individuals.

You should also realise that the enzyme, reverse transcriptase has practical application in genetic engineering as it enables us to make DNA copies (cDNA) of mRNA. We will not enlarge on this issue here since it is described in detail in the BIOTOL text, 'Techniques for Engineering Genes'.

SAQ 5.8

Using the information in Section 5.7 we now challenge you to modify the central dogma shown in 5.1.2.

Insert and label appropriate stages.

Summary and objectives

In this chapter we have outlined the major features of transcription in prokaryotic cells. Transcription of DNA into RNA depends upon the activity of RNA polymerase in recognising specific promoter sequences in the DNA. Production of a transcription unit of RNA can then be broken down into the key stages of initiation, elongation and termination. Our understanding of these processes has led to important new ideas in evolution and human health. Now that you have completed this chapter you should be able to:

- recognise experimental evidence for the existence of mRNA;

- describe the variety and levels of transcription products made by RNA polymerase;

- use supplied data to determine the rate of synthesis of RNA molecules;

- appreciate the implications of mRNA instability for transcription and translation;

- describe the structure and function of typical prokaryotic promoters;

- describe the sequence of events in the production of a transcription unit of RNA;

- recognise the role of mutants in dissecting the process of transcription;

- outline modifications of the central dogma concept of one-way flow of information and understand the importance of exceptions to this rule.

Translation

Translation

6.1 Introduction

In the previous chapter we have seen how RNA is produced in prokaryotes by the process of transcription. In this Chapter, the means by which the information (in the form of base sequence) within messenger RNA molecules is decoded and used to direct the synthesis of proteins will be described. This process is known as translation, and before discussing the details of the process itself, several concepts need to be addressed. The first of these is the nature of the genetic code carried by mRNA. As with any polynucleotide, the only variable part of the mRNA structure is the precise order of bases associated with the linear sugar-phosphate backbone, and indeed it is the nature and order of bases which dictates the sequence of amino acids in the encoded protein. The precise nature of this code and the ways in which it was determined will be discussed in Section 6.2.

The second important concept is the need for adaptor molecules through which amino acids are brought together as directed by mRNA. The requirement for, and the nature of, the adaptor molecules used (the tRNAs) are described in Section 6.3. Finally, an appropriate environment for the bringing together of amino acids is required. This is provided by ribonucleoprotein particles called ribosomes,which are described in detail in Section 6.4.

Translation involves the condensation of amino acids to form highly ordered protein structures, a process involving a decrease in entropy. To be thermodynamically feasible, therefore, a large input of Gibbs free energy is required. The energy source for translation comes from both GTP and ATP. The latter is involved in amino acid activation (Section 6.3) and the former in the initiation and elongation stages of translation itself (Section 6.5).

6.2 Elucidation of the genetic code

Once it was established that DNA is the genetic material of most organisms (exceptions being RNA viruses), and that the code was likely to reside in the order of bases along a polynucleotide chain, the scene was set for deducing the nature of the code. The mRNA, tRNA and rRNA produced by transcription are all required for protein synthesis, mRNA acting as the template to be decoded by the protein synthetic system. As RNA contains only four different bases (adenine, uracil, cytosine and guanine), and proteins contain twenty common amino acids, a single base: amino acid ratio for the code is not possible.

∏ What would be the possible number of different 'code words' (and hence amino
   acids) specified by combinations of 2, 3 and 4 bases?

The answers are :

for 2 bases, the number of combinations are $4 \times 4 = 4^2 = 16$;

for 3 bases, the number of combinations are $4 \times 4 \times 4 = 4^3 = 64$;

for 4 bases, the number of combinations are $4 \times 4 \times 4 \times 4 = 4^4 = 256$.

In other words, the number of combinations equals the number of different bases
available raised to the number of bases involved per code.

As we can see, combinations of 2 bases would give insufficient code words or codons
to specify 20 amino acids. The minimum number of bases involved to allow
independent specification of each amino acid is 3. A code based on triplet codons was
therefore proposed. Experimental proof that this is in fact the correct basis for the
genetic code was established in the 1960s by Crick and his co-workers. Their work on
viral genomes also indicated that the code is non-overlapping (Figure 6.1). As we can
see, the effect of a single base substitution would be more dramatic, affecting three
codons, if the code was overlapping. In fact, experimental evidence shows that such
base changes only influence single amino acids in the encoded proteins.

the code in
continuous and
non-overlapping

Another feature of the code is that it is continuous ie all the bases in a sequence of
mRNA are involved in coding, none are missed out. The effect of this is that the code is
read from a fixed starting point in a sequential manner.

∏ What will be the effect of introducing or deleting a base from the middle of an
   mRNA coding sequence?

frame-shift
mutations

Well, it should be obvious that the effect on a sequentially read, non-overlapping, code
will be that the codons after the insertion/deletion site will all be changed or 'shifted'
to specify a completely different sequence of amino acids. This is illustrated in
Figure 6.2, and such changes in reading frame of the code are known as frame-shift
mutations.

a) possible reading frames for a mRNA sequence starting from a fixed point

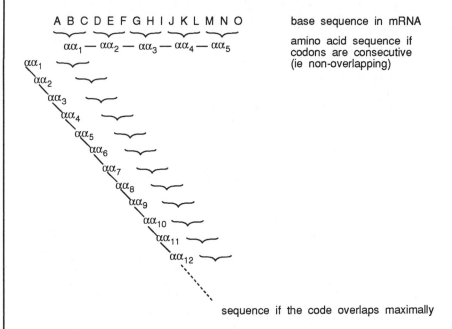

b) effect of changing a base in the mRNA

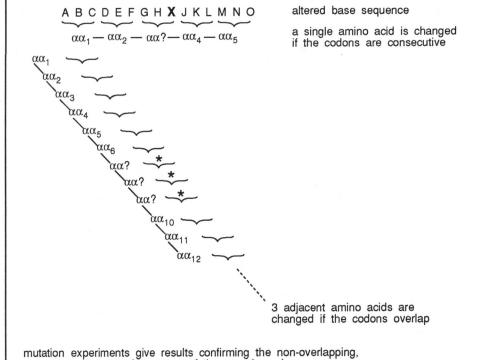

mutation experiments give results confirming the non-overlapping, consecutive nature of the genetic code.

Figure 6.1 Non-overlapping nature of the genetic code. $\alpha\alpha$ = amino acid.

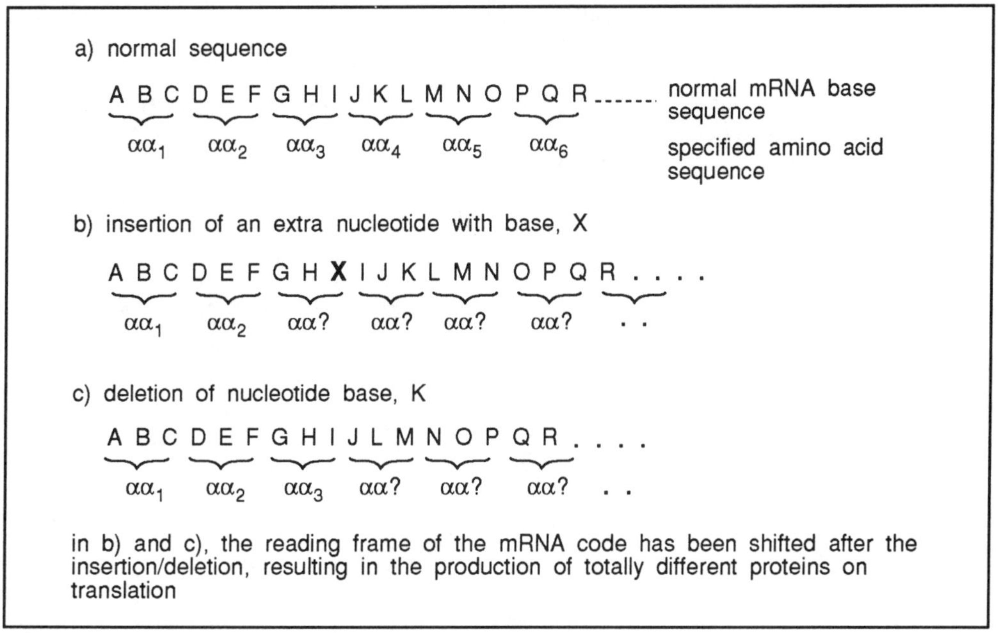

---

a) normal sequence

A B C D E F G H I J K L M N O P Q R ------ normal mRNA base
                                                       sequence

$\alpha\alpha_1$ $\alpha\alpha_2$ $\alpha\alpha_3$ $\alpha\alpha_4$ $\alpha\alpha_5$ $\alpha\alpha_6$ specified amino acid
                                                       sequence

b) insertion of an extra nucleotide with base, X

A B C D E F G H **X** I J K L M N O P Q R . . . .

$\alpha\alpha_1$ $\alpha\alpha_2$ $\alpha\alpha?$ $\alpha\alpha?$ $\alpha\alpha?$ $\alpha\alpha?$ . .

c) deletion of nucleotide base, K

A B C D E F G H I J L M N O P Q R . . . .

$\alpha\alpha_1$ $\alpha\alpha_2$ $\alpha\alpha_3$ $\alpha\alpha?$ $\alpha\alpha?$ $\alpha\alpha?$ . .

in b) and c), the reading frame of the mRNA code has been shifted after the insertion/deletion, resulting in the production of totally different proteins on translation

Figure 6.2 Effects of insertions/deletions on mRNA code.

The next problem which needed to be addressed was which of the 64 possible triplet codons coded for which amino acids. This deciphering of the code required a large amount of work, but was accomplished mainly as a result of three sets of experiments. The first of these was carried out by Nirenberg and Matthaei. They found that by adding a synthetic RNA to a cell-free protein - synthesising extract they could cause it to be translated. The synthetic mRNA they chose contained only uridine nucleotides (poly U), and was made using polynucleotide phosphorylase. This bacterial enzyme catalyses the formation of polyribonucleotides using 5′ nucleoside diphosphates (NDPs) as substrate:

$$NDP + (NMP)_n \rightarrow (NMP)_{n+1} + Pi$$

existing
polynucleotide

where N = U, C, G or A; Pi = inorganic phosphate; NDP = nucleoside diphosphate; NMP = nucleoside monophosphate.

The enzyme does not require a template, and thus catalyses the synthesis of random polynucleotides depending on the NDPs supplied. The availability of this enzyme in the late 1950s/early 1960s was essential to the success of these experiments as there was no other easy, reliable way to synthesise polyribonucleotides at this time.

RNA synthesis using a cell free extract and synthetic polynucleotide

The cell-free extract to which Nirenberg and Matthaei added their poly U molecules contained all the amino acids, ATP, tRNA, the enzymes responsible for linking amino acids to their specific tRNAs, messenger-depleted *E.coli* ribosomes etc. The result was the synthesis of a peptide containing only phenylalanine residues (poly-phe), and hence UUU was determined as a codon for this amino acid. Similarly, they found that CCC encodes proline, and other workers using the same technique found that AAA codes for lysine. It was not possible, unfortunately, to repeat this using a poly G template, as such molecules aggregate to form triple helices not readable by ribosomes.

By extending this approach to the use of synthetic RNA templates produced using different ratios of the ribonucleoside diphosphates with polynucleotide phosphorylase, Nirenberg and another group led by Ochoa deduced the bases involved in about 50 more codons specifying amino acids. As an example, when they produced synthetic mRNA using a molar ratio of UDP:GDP of 5:1, this gave RNAs with the following relative distributions of the different codons:

Possible triplets	Calculated frequency of occurrence relative to UUU (=100)
UUU	100
UUG	20
UGU	20
GUU	20
UGG	4
GUG	4
GGU	4
GGG	0.8

By comparing this to the relative amounts of amino acids inserted into the encoded peptides, identities of some codons could be assigned. In this particular experiment, the approximate relative amount of amino acids incorporated were:

phenylalanine	100
cysteine	20
valine	20
glycine	4
tryptophan	5

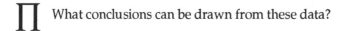 What conclusions can be drawn from these data?

By comparison with the table of expected codon frequencies, it should be obvious that UUU codes for phenylalanine (which we already know), that cysteine and valine are specified by codons containing two Us and a G, and glycine and tryptophan by codons containing two Gs and a U. We cannot, however, conclusively say which precise order the bases are in these codons from these experiments. The breakthrough which allowed more precise codon assignment came with the subsequent discovery by Nirenberg and Leder that in the presence of 20mmol l^{-1} MgCl$_2$ and absence of GTP, $E.coli$ ribosomes will bind synthetic RNAs as short as 3 nucleotides in length provided they are phosphorylated at the 5' end. In addition, the ribosomes will bind the adaptor tRNA carrying the amino acid specified by the triplet codon on the trinucleotide eg pUUU will cause phenylalanyl-tRNA to be bound.

ribosome
binding assay

This discovery enabled the development of an experimental strategy for the determination of amino acids coded by most of the codons. Basically, a cell-free system such as described previously was used, with the Mg^{2+} concentration adjusted to 20mmol l^{-1}, and this was added to a series of tubes each containing the same nucleotide triplet and a different radioactively labelled amino acid. Following 37°C incubation to allow ribosome binding, samples were passed through cellulose nitrate filters. Ribosomes and compounds associated with them were retained on the filter and any unbound amino-acyl tRNAs passed through them (see Figure 6.3). By monitoring retention of radioactively labelled amino acids on the filter, it was hence possible to determine exactly for which amino acid approximately 50 of the possible 64 codons specified. The remaining triplets either gave equivocal results or failed to induce binding of any amino-acyl tRNAs.

Examine Figure 6.3 carefully. You will see in stage 1) that ribosomes, tRNAs, one radioactively labelled and nineteen unlabelled amino acids, a trinucleotide and the appropriate enzymes are incubated at 37°C. During this incubation the amino acids become attached to their appropriate tRNAs. Then an amino acid-tRNA forms hydrogen bonds with the trinucleotide associated with the ribosome to form a structure like that shown at the bottom of the figure.

In part 2)a of Figure 6.3, the amino acid specified by the mRNA triplet on the trinucleotide is one of the 19 unlabelled amino acids. Hence, the radioactive amino acid bound to its tRNA passes through the cellulose nitrite filter. In part 2)b of Figure 6.3, the labelled amino acid is that specified by the trinucleotide base sequence. It is, therefore, retained on the filter in the form of aminoacyle-tRNA bound to ribosomes.

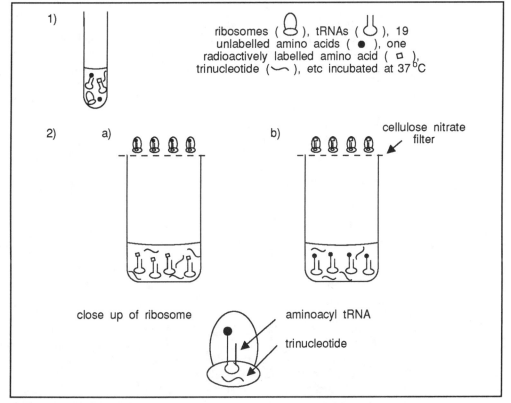

Figure 6.3 Ribosome binding assay for the assignment of triplet codons to individual amino acids. (See text for a description of this assay).

To determine the assignments for the remaining codons, further experiments using longer synthetic RNAs were carried out by Khorana. He made polyribonucleotides with repeated defined sequences by transcribing synthetic DNA with RNA polymerase. The synthetic DNA produced contained repeated sequences involving two or three of the deoxynucleotides in each strand, eg:

                5' T A C T A C T A C T A C T A C 3'
                3' A T G A T G A T G A T G A T G 5'

Selective transcription of one of the strands by RNA polymerase is possible by adding 3 of the 4 ribonucleotides.

∏ What will be the RNA sequence produced on adding GTP, ATP and UTP?

As written, it will be the top strand of the DNA which acts as the template for formation of an RNA with these nucleotides, so starting from the 5' end, the transcript will be:

5' GUAGUAGUAGUAGUA 3' (Note that the DNA is transcribed from the 3' → 5' end). We will call this transcript a.

Alternatively, if supplied with CTP, ATP and UTP, the transcript produced would be:

5' UACUACUACUACUAC 3'. This we will call transcript b.

When supplied individually to a cell-free protein synthesising extract, each of these mRNAs has the potential to give rise to 3 polypeptides, each of which will contain only one type of amino acid. For example, transcript (a) above has three possible reading frames depending on whether the first codon recognised starts with a G (giving multiple GUA codons), U (giving UAG) or A (AGU). If the amino acids coded by one or two of these is already known, then by analysing the peptides produced, the amino acid encoded by the third codon can be determined. In actual fact, in the example used, poly (GUA), only two homopolymeric peptides were coded, as one of the possible codons, UAG, does not code for an amino acid, but is used to terminate polypeptide synthesis ie it is a 'stop codon'. We will return to this concept later.

The combined data from these experiments completely account for the 64 codons of the genetic code. The amino acid assignments are shown in Table 6.1. The most strikingly obvious thing about this table is that most amino acids are coded by more than one codon, such that 61 codons specify 20 amino acids. In other words, the code is degenerate, although it is not in any way ambiguous - each codon specifies a single amino acid or causes termination of protein synthesis. Different codons which specify the same amino acid are called synonyms eg CUU and CUG are synonymous codons for leucine. Generally, most synonyms have the same first two bases, and differ at the third position. In particular, codons with the same first and second base all code for the same amino acids if their third bases are U and C (pyrimidines), and usually code for the same amino acids if the third bases are A and G (purines). Only serine, leucine and arginine have six synonyms, and only methionine and tryptophan are coded by single codons. The codon for methionine is important as it is the one used in most circumstances to initiate protein synthesis, as we shall see in Section 6.5.1. Three codons do not specify amino acids, but serve an important function in causing termination of protein synthesis once sufficient amino acids have been polymerised. This will be elaborated on in Section 6.5.3.

		U		C		A		G	
U		UUU	Phe	UCU	Ser	UAU	Tyr	UGU	Cys	U
		UUC	Phe	UCC	Ser	UAC	Tyr	UGC	Cys	C
		UUA	Leu	UCA	Ser	UAA	**STOP**	UGA	**STOP**	A
		UUG	Leu	UCG	Ser	UAG	**STOP**	UGG	<u>Trp</u>	G
C		CUU	Leu	CCU	Pro	CAU	His	CGU	Arg	U
		CUC	Leu	CCC	Pro	CAC	His	CGC	Arg	C
		CUA	Leu	CGA	Pro	CAA	Gln	CGA	Arg	A
		CUG	Leu	CCG	Pro	CAG	Gln	CGG	Arg	G
A		AUU	Ile	ACU	Thr	AAU	Asn	AGU	Ser	U
		AUC	Ile	ACC	Thr	AAG	ASn	AGC	Ser	C
		AUA	Ile	ACA	Thr	AAA	Lys	AGA	Arg	A
		AUG	<u>Met</u>	ACG	Thr	AAG	Lys	AGG	Arg	G
G		GUU	Val	GCU	Ala	GAU	Asp	GGU	Gly	U
		GUC	Val	GCC	Ala	GAC	Asp	GGC	Gly	C
		GUA	Val	GCA	Ala	GAA	Glu	GGA	Gly	A
		GUG	Val	GCG	Ala	GAG	Glu	GGG	Gly	G

Table 6.1 The genetic code.

Three codons do not specify amino acids, but cause termination of polypeptide synthesis. These are shown in bold, and are UAA, UAG and UGA. Only two amino acids are specified by single codons, these are methionine (AUG) and tryptophan (UGG), and are shown underlined.

∏ What do you think the advantages of having a degenerate genetic code are?

code degeneracy

As the triplet nature of the code means that there are 64 possible codons to specify 20 amino acids, the use of 61 of them for this purpose has the effect of minimising damage due to genetic mutations. Due to mistakes in DNA replication, errors arise in copying a template strand at a frequency of approximately 1 in 10^8 bp. These errors usually take the form of substitution of an incorrect for a correct nucleotide. If this substitution occurs in a region of the genome coding for a protein, the fact that most mutations will give rise to an alternative amino acid-specifying codon means that although a slightly altered protein product may be produced this is unlikely to cause a major threat to the organism (although there are some notable exceptions where single base, and hence amino acid, substitutions produce lethal phenotypes). If the genetic code was not degenerate, this would mean that we would have 20 coding codons and 44 stop codons. Under this scenario the likelihood of a single mutation giving rise to premature polypeptide chain termination would be very high. In this case no functional protein would be produced at all, possibly causing severe phenotypic effects.

An additional advantage of code degeneracy, is that this degeneracy is not random - synonyms are closely related as we have seen, as are codons for amino acids with similar side chains. Again this minimises the effects of mutations, single DNA and hence mRNA base changes being likely to result in substitution of the correct amino acid or a closely related one.

One feature of the genetic code which so far we have taken for granted, is that it is almost universal, ie a human mRNA will be translated by a bacterial extract to produce the same polypeptide chain as that produced in a human cell. There are only a very few examples of different codon usage to that described in Table 6.1, indicating the centrally important role of this code and its ancient origin. The selective pressure during evolution to retain the same codon usage has obviously been very high. Some examples of alternative codon usage are shown in Table 6.2, and all occur in the mitochondrial genomes of eukaryotes. As you may recall, mitochondria have their own small circular DNA which is expressed to produce certain mitochondrial proteins. Mitochondria are thought to represent ancient prokaryotic organisms which set up symbiotic relationships with eukaryotes enabling the latter to take advantage of an aerobic environment. Hence the mitochondrial DNA may represent a very ancient 'genome' which used a slightly different code or which has been under less selective pressure to maintain the same code as the nuclear genome. The latter seems a more likely explanation, as the number of proteins or subunits coded by mitochondrial DNA is very small (about ten).

Codon	Standard code meaning	Meaning in mitochondria	Organism
UGA	STOP	Trp	yeast, invertebrates and vertebrates
AG(G/A)	Arg	STOP	mammals
AUA	Ile	Met (initiation)	mammals
		Met (initiation)	fruit fly
		Met(elongation)	yeast
CUA	Leu	Thr	yeast
AGA	Arg	Ser	fruit fly

Table 6.2 Alternative codon usage.Met (initiation) indicates that this codon acts as the start signal for protein synthesis. See Section 6.5.1 for details of translational initiation. Met (elongation) indicates that this codon specifies insertion of methionine into a growing polypeptide chain. AG(A/G) means AGA and AGG). Data taken from Lewin (1990), Genes IV, Oxford University Press, Oxford.

SAQ 6.1

1) Summarise the important features of the genetic code (try to list about 8 features).

2) If you knew that CUC codes for leucine, describe an experiment by which you could deduce the amino acid specified by UCU.

6.3 Transfer RNA - the adaptor molecule in protein synthesis

tRNA

amino-acyl tRNA synthetase

As mentioned in the introduction, adaptor molecules are required to align amino acids up on adjacent mRNA codons for protein synthesis. They are required as there is no physical or chemical way in which a codon triplet of nucleotide bases could itself recognise a small amino acid molecule. The adaptors take the form of tRNA molecules, and their particular amino acids are linked to them by highly specific enzymes, the amino-acyl tRNA synthetases. There are at least 20 different types of these enzymes per cell (one for each amino acid), and they vary markedly in structure as shown in Table 6.3.

Amino acid specified	Molecule weight (KD)	Subunit structure
histidine	85	$\sigma 2$
isoleucine	114	single chain
lysine	104	$\sigma 2$
glycine	227	$\sigma \ \beta 2$

Table 6.3 Properties of various *E.coli* amino-acyl tRNA synthetases. The enzymes from other species are similarly diverse.

Linking an amino acid to its tRNA has the effect of activating it for participation in protein synthesis. This activation requires an input of energy, and in fact two high energy phosphate bonds are invested in the formation of each amino-acyl tRNA. The reaction scheme is shown in Figure 6.4.

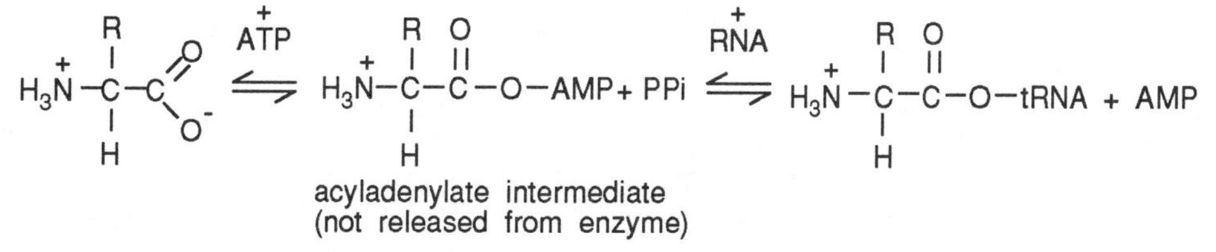

Figure 6.4 Activation of amino acids and their attachment to tRNA.

The reaction catalysed by amino-acyl tRNA synthetase is readily reversible. However, hydrolysis of the pyrophosphate (PPi) released from ATP, by pyrophosphatase released sufficient free energy to pull the reaction towards completion. The high specificity of the enzyme for the amino acid and the tRNA to which it is joined is essential for reliable protein synthesis.

We now need to consider the nature of tRNAs and the ways in which they recognise RNA codons and bind amino acids. The first tRNA was sequenced in 1965 by Holley, and was the yeast tRNA for alanine - tRNA[ala]. This sequence yielded many of the important features of tRNAs and paved the way for sequencing of many more.

2D: clover-leaf
structure of
tRNA

In general, tRNAs are similar in size (73 to 93 nucleotides long), and in secondary and tertiary structure. Additionally, they contain a high proportion of unusual bases (examples in Figure 6.5), usually start at the 5′ end with a guanine nucleotide and always end at the 3′ terminus with CCA-OH. It is to the 3′ adenosine that the appropriate amino acid is linked, although linkage may be to either the 2′ or 3′ hydroxyl group. The secondary structure adopted by tRNAs takes the form of a clover-leaf pattern due to extensive intra-strand basepairing as shown in Figure 6.6. The five 'arms' produced have been given names reflecting their base composition or function. The 5′ and 3′ ends of the molecule basepair to produce the amino acid acceptor arm, with the 3′ terminal CCA sequence protruding as the amino acid binding site. The DHU arm is named for the occurrence in it at variable positions of dihydrouridine, and the TψC arm for the presence of the base sequence: ribothymine-pseudouracil-cytidine. The

anticodon

anticodon arm contains the three nucleotide anticodon sequence in the unpaired loop at its end. This sequence, as the name suggests, is complementary to the mRNA codon specifying the amino acid carried on the amino acid acceptor arm. Between the TψC and the anticodon arms is found the variable arm, the length of which depends on the tRNA.

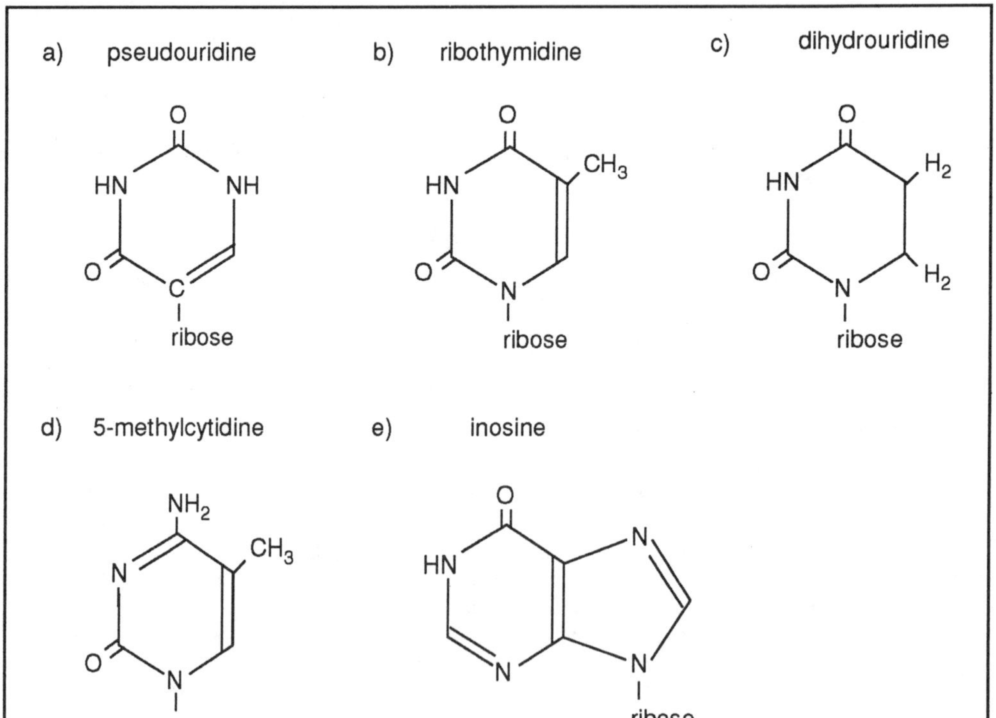

Figure 6.5 Some unusual nucleotides found in tRNA molecules.

3D: L-shaped
tRNA

It is easy to draw tRNA molecules in the two dimensional clover-leaf form, but they are in reality three dimensional, and adopt an L-shaped tertiary structure due to unusual hydrogen-bonding involving not only bases which do not normally basepair, but elements of the sugar-phosphate backbone also. The overall effect is that the amino acid attachment point is at one end of the L, and the anticodon at the other, with the DHU and TψC loops forming the corner between the two.

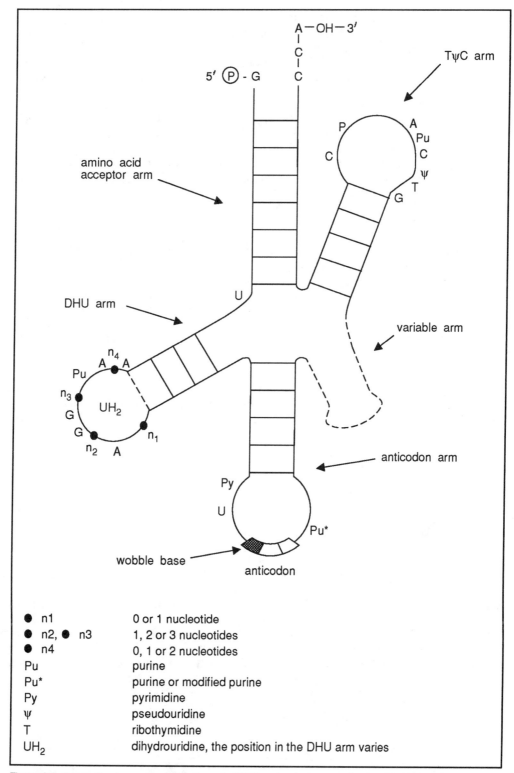

Figure 6.6 Generalised secondary structure of a tRNA molecule (see text for details). Note that we have not attempted to put all of the modifications of the basic structures of tRNA on this figure.

You should note that the unusual nucleotides of tRNAs contain derivatives of the usual bases, produced by enzymatic action on precursor tRNAs. Many modified bases are methylated; this has the dual effect of altering their basepairing capacities and making them hydrophobic. This latter property might be important in allowing tRNAs to interact with other components of the protein synthetic machinery.

wobble hypothesis

As we have seen, many amino acids are coded by synonyms. Interestingly, this does not mean that the cell must produce a separate tRNA to recognise each possible codon for a particular amino acid. Some tRNAs are able to recognise two or even three synonymous codons. They can do this due to the fact that basepairing between the 5' base of the tRNA anticodon (the first base of the anticodon) and the 3' base in the mRNA codon (the third base of the codon) exhibits a degree of flexibility or 'wobble'. The 'wobble hypothesis' (which states that one tRNA can recognise more that one codon due to flexibility in basepairing with the third base in certain codons) was put forward by Crick and explained some of the degeneracy in the genetic code. As the first two bases in a codon basepair in the ordinary way, these must (as we saw in Figure 6.1) specify a particular amino acid and hence codons which differ at either of these positions must be recognised by distinct tRNAs. The possibility for variation in third codon base basepairing explains the fact that the unusual base inosine is often observed as the first base in anti-codons. Inosine (see Figure 6.7b) is a very flexible base in that when in the wobble position it is capable of basepairing with adenine, uracil or cytosine. Similarly, under the less stringent conditions in the wobble position, guanine can basepair with uracil in addition to cytosine, and uracil with guanine or adenine (see Figure 6.7).

SAQ 6.2

Using the information described above and in Table 6.1a, what are the minimum numbers of tRNAs required to recognise all the synonyms of:

1) phenylalanine;

2) isoleucine;

3) serine;

4) proline?

initiator tRNA

tRNA_f and tRNA^met

When we come to a discussion of initiation of protein synthesis (Section 6.5.1) we shall see that initiation in prokaryotes almost always occurs at a specific AUG codon near to the 5' end of the mRNA. This codon is recognised by a special tRNA known as the initiator tRNA, or tRNA_f and specifies a modified methionyl residue. The first residue of all peptides is N-formyl methionine (specified by tRNA_f) and is illustrated in Figure 6.8. Its production occurs in two stages; firstly the linkage of a methionyl residue to the tRNA, and secondly the N-formylation of this residue. The first reaction is carried out by methionyl-tRNA synthetase (which also joins methionyl residues to tRNA^met for insertion into polypeptides at other positions), and the second by transformylase, which transfers a formyl group from N^{10}-formyl-tetrahydrofolate.

1) possible basepairs at the wobble position

5' base of anticodon	corresponding 3' base of codon
G	C or U
C	G
A	U
U	A or G
I	A, U or C

I = inosine

2) hydrogen bonding involved in unusual basepair arrangements

a) guanine-uracil

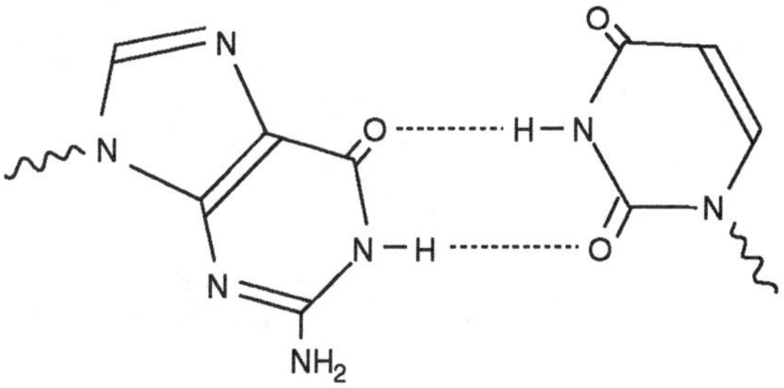

b) inosine-adenine

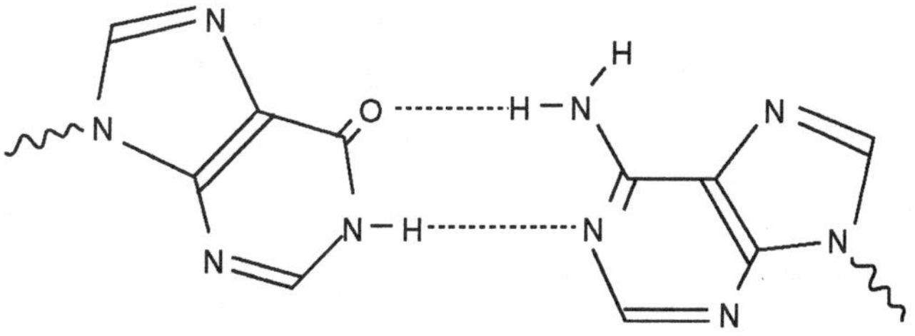

c) inosine-uracil

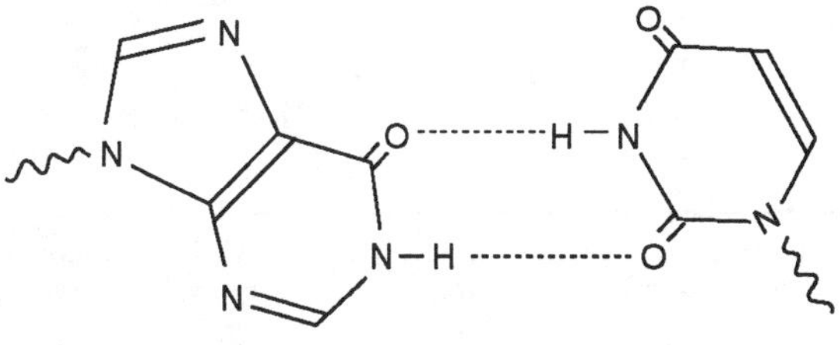

d) inosine-cytosine

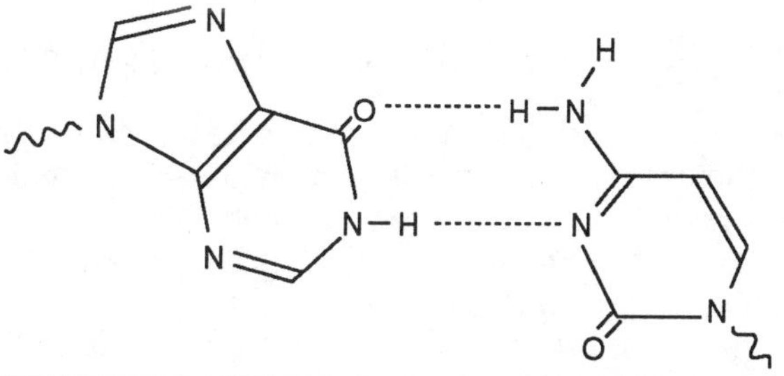

Figure 6.7 Basepairing observed between the 3'-most residues of codons and the 5'-most residues of anticodons: the wobble hypothesis.

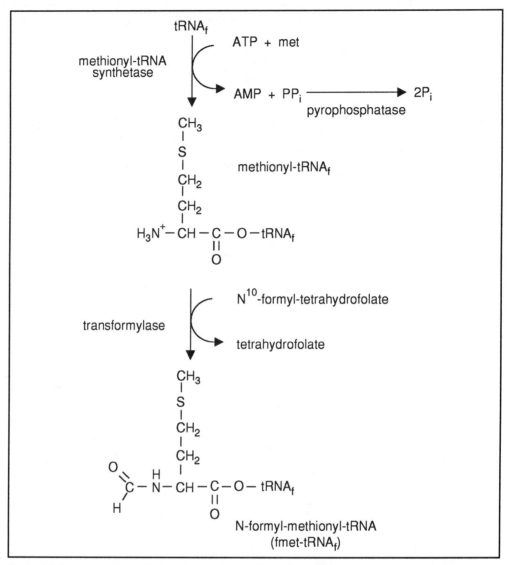

Figure 6.8 Formation of the initiation formyl-methionyl tRNA. Note that the formyl group or the whole formyl-methionyl residue may be removed from the protein post-translationally to produce the mature amino terminus.

6.4 Ribosomes - the site of protein synthesis

rRNA   Ribosomes are complex ribonucleoprotein particles - aggregates of RNA and protein - that provide the environment for protein synthesis, in which amino-acyl tRNAs are brought together by the mRNA base triplet code. All ribosomes (approximately 20000) within a particular bacterium are identical in structure, and able to drive translation of any of that cell's mRNA. The best characterised ribosomes are those of *E.coli*. Generally, ribosome components are described in terms of their sedimentation coefficients in Svedberg(S) units. This value relates to the movement of an entity through a solution under a centrifugal field and although related to the entity's mass it also takes into account its shape. S values are not additive. The intact ribosome has a sedimentation coefficient of 70S, a mass of 2700kD and a diameter of approximately 20nm. It consists of two subunits, a small 30S subunit and large 50S one (Figure 6.9). Unless actively engaged in translation, the ribosome dissociates into its individual subunits (see Section 6.5), each of which comprises both RNA (ribosomal RNA or rRNA) and protein. The small subunit contains a single 16S rRNA molecule and one copy of each of 21 proteins (S1-S21). The nature and functions of these constituents and those of the large subunits (5S and 23S rRNAs and 34 proteins, L1-L34) are shown in Tables 6.4 and 6.5, and some of these functions will be further explained in the next section. There is a lot of information in Tables 6.4 and 6.5. At this stage, just read through these tables. You will realise the significance of the function of the ribosomal proteins when you have read the next section.

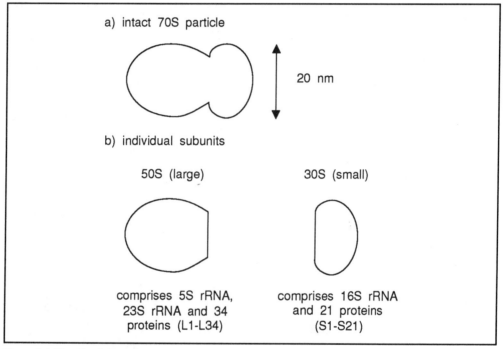

Figure 6.9 The *E.coli* ribosome.

Protein	No amino acid residues	Function
S1	557	mRNA binding
S2	240	mRNA binding
S3	232	mRNA binding; tRNA binding?
S4	203	30S subunit folding
S5	166	mRNA binding; tRNA binding
S6	131	-
S7	177	30S subunit folding
S8	129	30S subunit folding
S9	128	mRNA binding; tRNA binding?
S10	103	mRNA binding?; tRNA binding
S11	128	-
S12	123	tRNA binding; mRNA binding
S13	117	-
S14	98	tRNA binding; mRNA binding?
S15	87	30S folding
S16	82	30S folding
S17	83	30S folding
S18	74	-
S19	91	tRNA binding; mRNA binding
S20 = L26	86	-
S21	70	mRNA binding
L1	233	part of A site
L2	272	peptidyl transfer; part of P site
L3	209	peptidyl transfer; 50S subunit folding
L4	201	As L3
L5	178	peptidyl transfer?; part of A site
L6	176	peptidyl transfer
L7/L12	120	protein factor binding; translocation; part of A site
L9	148	-
L10	163	-
L11	141	peptidyl transfer
L13	142	-
L14	120	part of P site
L15	144	peptidyl transfer

Table 6.4 continued

L16	136	peptidyl transfer
L17	127	-
L18	117	peptidyl transfer
L19	114	-
L20	117	50S subunit folding; part of A site
L21	103	-
L22	110	-
L23	99	-
L24	103	50S folding; part of P site
L25	94	peptidyl transfer?
L27	84	part of P site; peptidyl transfer?
L28	77	-
L29	63	-
L30	58	part of A site
L31	62	-
L32	56	-
L33	54	part of P site
L34	46	-

Table 6.4 Properties and functions of *E.coli* ribosomal proteins.- indicates function unkown. ? indicates that this function has not been completely proven. L7 and L12 are alternate forms of the same protein. L7 has an acetylated N-terminal end, whereas L12 does not. Two copies of each peptide are found per ribosome, and aggregate to form a functional tetramer. S20 and L26 represent the same protein found in both ribosomal subunits. Data from Matthews and Van Holde (1990), Biochemistry, Benjamin-Cummings Press, UK.

rRNA	No. nucleotides	Function
16S	1542	Shine-Delgarno sequence recognition, small subunit integrity, part of P and A sites
5S	120	maintenance of large subunit integrity
23S	2900	as 5S, part of A site, peptidyl transfer, binds deacylated tRNA

Table 6.5 The nature and functions of *E.coli* rRNAs.

In terms of mass, the rRNAs make up almost two thirds of ribosomes. They are highly folded due to intrastrand basepairing, and are essential for the self-assembly of large and small subunits from their individual components which can be studied *in vitro* and hence probably also occurs *in vivo* without the requirement of additional protein factors.

The use of electron microscopy, specific antibodies, and neutron diffraction has allowed the overall shape of ribosomes and the arrangements of proteins within them to be mapped to a high degree of certainty. The overall shapes of large and small subunits and of the intact ribosome are shown in Figure 6.10. The mRNA binding site is located on the platform of the 30S subunit, and the two tRNA binding sites of the ribosome are in the cleft between this platform and the rest of the subunit. The enzyme catalysing peptide bond formation, peptidyl transferase, is located in the valley between two of the protruberances on the large subunit. The thinnest protruberance contains the GTPase activity (catalysing GTP → GDP and Pi) which provides energy for mRNA and tRNA movements. As protein is synthesised, the growing chain leaves the ribosome through the opposite edge. A simplified diagram of the important sites is shown in Figure 6.11.

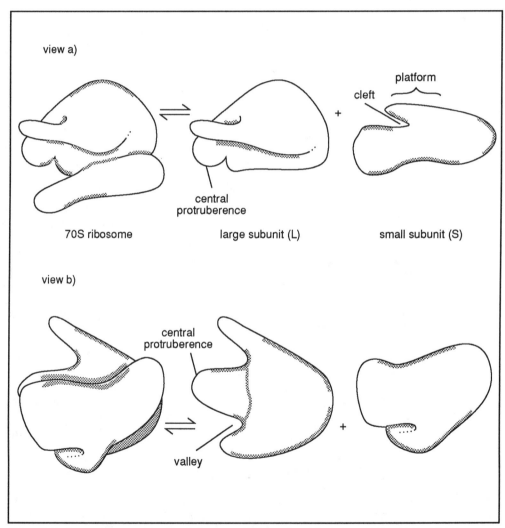

Figure 6.10 Gross structure of the *E.coli* ribosome. Redrawn from Matthews and Van Holde (1990), Biochemistry, Benjamin-Cummings Press.

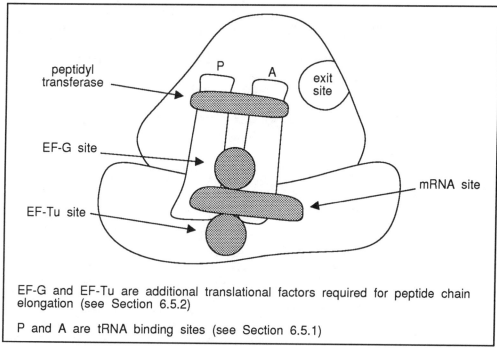

EF-G and EF-Tu are additional translational factors required for peptide chain elongation (see Section 6.5.2)

P and A are tRNA binding sites (see Section 6.5.1)

Figure 6.11 Simplified diagram to show important regions of the *E.coli* ribosome. Adapted from Lewin B (1990), Genes IV, Oxford University Press, p 167.

6.5 The process of translation

Now that we have discussed the important ingredients for translation - the genetic code, tRNA adaptor molecules and ribosomes - we can consider the process itself.

∏ As is the case for transcription, translation occurs in three stages - what are they?

The three stages we hoped you would get are initiation (starting the process), elongation and termination of the completed polypeptide. As for transcription these stages all have specific requirements, and we shall look at them in turn. Before this, however, we will consider an important facet of bacterial gene expression, that is the processes of transcription and translation are very closely coupled. As the DNA is not 'locked away' in a nucleus, translation of mRNA molecules can commence before their completion. In other words, the processes are coupled temporarily (ie in time) as well as in space within bacteria. This is important for bacteria to allow sufficient protein synthesis as they have such a short time between cell divisions. Additionally, control mechanisms for regulation of gene expression have evolved to make use of this property. These mechanisms include attenuation, which is the subject of Chapter 8 of this text.

∏ What does the close coupling between transcription and translation tell us about translation?

Well, as mRNA is synthesised in the $5' \rightarrow 3'$ direction, if translation is to start prior to mRNA completion, it must occur by reading the mRNA in the $5' \rightarrow 3'$ direction. In our discussions of the genetic code, we have already made this assumption, but it is important that you should realise it as fact! An additional point for which there is ample experimental evidence is that proteins are synthesised in the $N \rightarrow C$ direction ie starting at the amino terminal amino acid and finishing at the carboxyl end.

Now let us consider the stages of translation.

6.5.1 Initiation

initiation factors

To start translation, mRNA, fmet-tRNA$_f$, small and large ribosomal subunits and a second amino-acyl tRNA must be brought together. This requires the additional activities of proteins called initiation factors (IF), and occurs in a co-ordinated fashion (see Figure 6.12). Use this figure to follow the description given below.

Shine-Delgaron sequence

When not engaged in protein synthesis, ribosomal 30S subunits are prevented from association with 50S subunits by interaction with the initiation factor IF3. Before association with mRNA can occur, the 30S subunit forms a complex with two further IFs, IF1 and IF2, the latter of which has a molecule of GTP bound to it. IF interaction activates the 30S subunit such that mRNA can now interact with it. We mentioned earlier that, in bacteria, translation almost always starts at a special methionine codon - the initiation codon, AUG. This is recognised as the initiation codon as it is found approximately 10 nucleotides downstream of (in the 3' direction from) a purine rich region known as the Shine-Delgarno sequence. In rare instances, the GUG codon is used to initiate translation. When used, this codon is also recognised by fmet-tRNA$_f$. Initial contact between the mRNA and 30S subunit occurs through complementary basepairing between the Shine-Delgarno sequence and the 16S mRNA. Usually this interaction involves 3-6 complementary nucleotides. Once mRNA has bound, the initiation codon is positioned such that, under the influence of IF2, fmet-tRNA$_f$ is attracted to it. The initiation AUG occupies one of the two codon sites in the ribosome - called the P site. It is called this because as we shall see, as translation continues, this site is occupied by the tRNA carrying the growing peptide chain. The structure formed at this stage is known as the 30S initiation complex.

Subsequently, the 50S subunit is able to interact with the 30S initiation complex as IF3 has been removed. 50S subunit interaction causes hydrolysis of the GTP associated with IF2, hence displacing IF2 and IF1. The initiation complex is now complete, with the initiation tRNA$_f$ bound to the P site. The second amino-acyl tRNA binding site is vacant, awaiting the appropriate incoming molecule. This is called the A site as it binds the amino-acyl tRNA molecule specified by the mRNA codon positioned there.

SAQ 6.3

We have assumed so far that amino-acyl tRNA codon recognition occurs through the anticodon. How might you prove this experimentally?

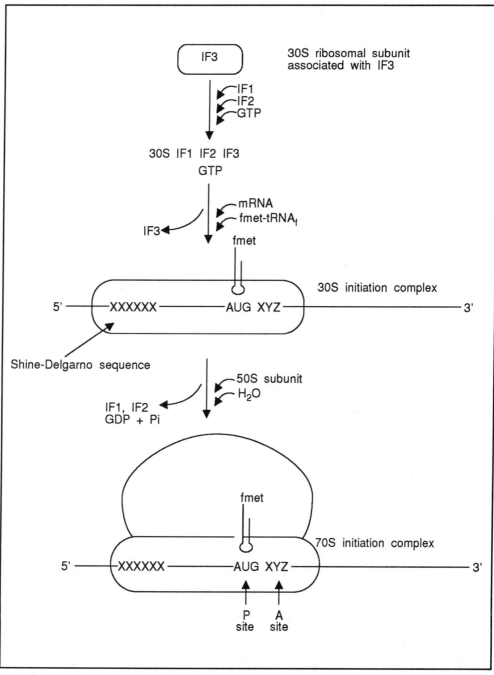

Figure 6.12 Initiation of *E.coli* protein synthesis (stylised). The ribosome contains two tRNA binding sites: the P site, which binds the initiating tRNA, and the A site, which is ready to accept the aminoacyl-tRNA specified by the codon XYZ. The first step involves the free 30S subunit (associated with IF3) interacting with two additional initiation factors (IF1 and IF2) and GTP. This complex allows the interaction of mRNA via 16S rRNA-Shine-Delgarno sequence basepairing. IF2-GTP attracts fmet-tRNAf to the initiation codon, AUG, and the 30S complex is produced. Subsequent interaction with the 50S subunit and GTP hydrolysis (involving L7/L12) releases IF1 and IF2, producing the mature initiation complex.

6.5.2 Elongation

On formation of the 70S initiation complex (Figure 6.12), the stage is set for peptide chain formation. This process again involves additional protein factors called elongation factors (EF). There are three of these, EF-Tu, EF-Ts and EF-G. EF-Tu is required to bring the amino-acyl tRNA specified by the A site codon to the ribosome. EF-Tu is a GTP binding protein, and it is the GTP-bound form which introduces amino-acyl tRNA to the A site (See Figure 6.13a). The factor has inherent GTPase activity and slowly hydrolyses its bound GTP to release inorganic phosphate. Following a short pause (several milliseconds) after hydrolysis, EF-Tu-GDP leaves the A site and the amino-acyl tRNA remains there. A peptide bond can only be formed once EF-Tu has left the A site.

elongation factors (margin note)

Π   The fact that EF-Tu hydrolyses the bound GTP slowly, and subsequently pauses before leaving the A site is very important. See if you can think of an explanation for this before reading on.

As peptide bond between the existing peptide chain on the tRNA in the P site and the incoming residue on the tRNA at the A site is formed only after EF-Tu has left, this gives the ribosome a chance to 'check' that the correct amino-acyl tRNA has entered. It effectively does this by scanning the codon-anticodon basepairing. It is possible for a tRNA having only 2 of the 3 anticodon bases pairing to the mRNA codon to enter the ribosome, and if no checking was carried out the wrong amino acid could be introduced into the peptide chain. The intervals before GTP hydrolysis, and following hydrolysis but before EF-Tu dissociation from the ribosome allow time for incorrect tRNAs to leave the A site and the correct one to enter. Additionally, EF-Tu changes conformation on GTP hydrolysis, altering the codon-anticodon context (see Figure 6.11). This effectively means that tRNA-mRNA basepairing is checked twice. Only the correct amino-acyl tRNA will interact with the codon under both contextual conditions.

checking codon anticodon basepairing (margin note)

The overall effect of this scrutinisation of the incoming amino-acyl tRNA is that the error rate for protein synthesis is maintained at about 10^{-4} per amino acid. Increasing the pause time would lower this rate further. However, the value achieved is optimal to the cell as it allows protein synthesis to occur at a relatively fast rate with a very high proportion of error-free products. Increasing the rate of EF-Tu GTP hydrolysis would result in the production of too many incorrect proteins.

Once the correct amino-acyl tRNA has associated with the A site, peptidyl transferase catalyses the transfer of the amino acid (or the peptide chain) associated with the tRNA at the P site onto that on the A site (Figure 6.13). The enzyme activity is located on the large ribosomal subunit.

Π   Briefly sketch the type of reaction involved in formation of the peptide bond between fmet and a second aminoacyl-tRNA (do not worry about the amino acid side chains). Remember that fmet is esterified to tRNA$_f$.

You should have come with something like the reaction shown in Figure 6.13b. This is thermodynamically favourable because two high energy phosphate bonds were used in activating the fmet by linking it to the tRNA$_f$. Read the legend to Figure 6.13 carefully as it provides a useful summary.

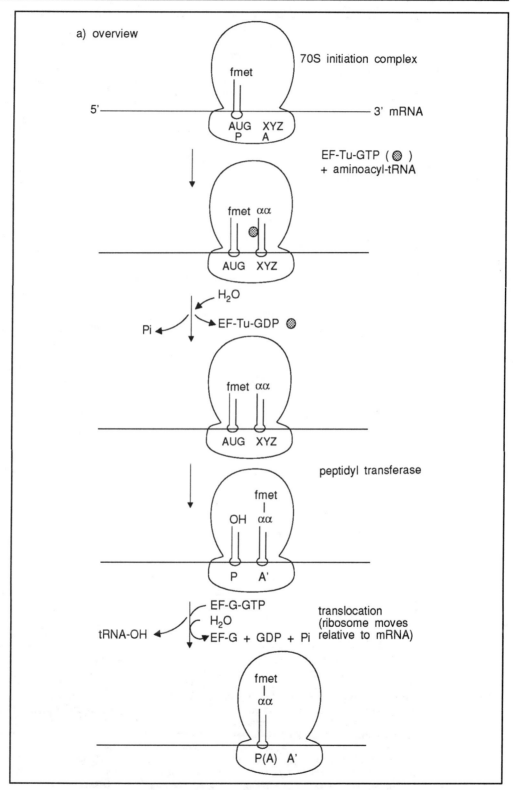

Figure 6.13 a) Overview of the elongation cycle in *E.coli* protein synthesis. See legend at the end of the figure on the next page.

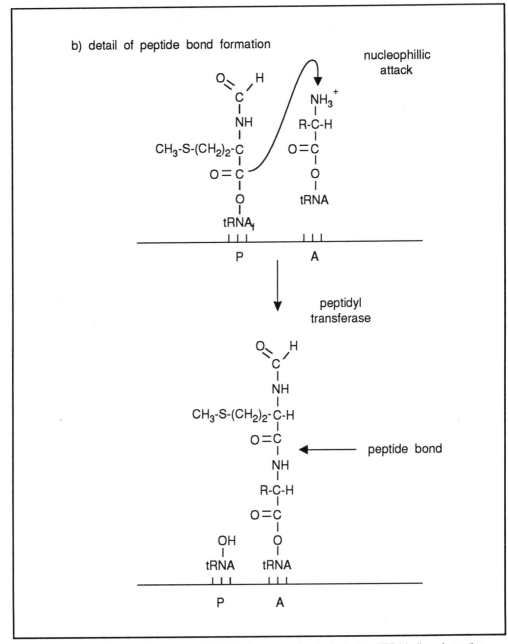

Figure 6.13 a) The appropriate amino-acyl tRNA is brought to the A site by the GTP-binding elongation factor EF-Tu. After a pause, the GTP associated with EF-Tu is hydrolysed by the endogenous GTPase activity of the factor. EF-Tu·GDP is released, and the aminoacyl tRNA at the A site is committed to participation in translation. Peptidyl transferase (an enzyme activity of the large ribosomal subunit) catalyses the transer of fmet onto the amino-acyl moiety of the tRNA occupying the A site, with the formation of a peptide bond. The uncharged tRNA then leaves the P site as GTP is hydrolysed and the ribosome moves along the mRNA in the 5'-3' direction. This process is called translocation, and positions the peptidyl-tRNA in the P site, leaving the A site occupied by the next triplet codon and available for amino-acyl tRNA to enter. b) Peptidyl transferase catalyses a nucleophillic attack by the fmet carbonyl group on the alpha amino group of the amino-acyl residue in the A site. The result is that a peptide bond is formed with the amino acyl residue in the A site and a free tRNA is left in the P site.

Following peptide bond formation, the P site is occupied by an uncharged tRNA. This must leave the ribosome, which in turn must move along the mRNA such that the peptidyl-tRNA enters the P site and the next codon enters the A site. These processes are co-ordinated and collectively termed translocation. Translocation requires a further elongation factor EF-G (or translocase). This is another GTP binding protein which cycles between a GTP and a GDP form. EF-G-GTP binds to the ribosome (Figure 6.11) to drive translocation by GTP hydrolysis. Following translocation EF-G-GDP leaves the ribosome, and the A site is available for a new incoming amino-acyl tRNA. The process of amino-acyl tRNA binding, peptide bond formation and translocation is repeated over and over until a stop codon enters the A site, and termination of polypeptide synthesis occurs.

One point which we must address before considering translational termination is how EF-Tu-GTP is recycled. On GTP hydrolysis, GDP remains bound to EF-Tu, and before the factor can further participate in translation, this GDP must be replaced by GTP. This recycling involves a further factor, EF-Ts, and is illustrated in Figure 6.14. EF-Ts binds to EF-Tu-GDP allowing replacement of bound GDP by GTP. Once EF-Tu-GTP is formed, EF-Ts dissociates, and EF-Tu - GTP is available for use in transporting amino-acyl tRNA to the ribosome A site for peptide chain elongation.

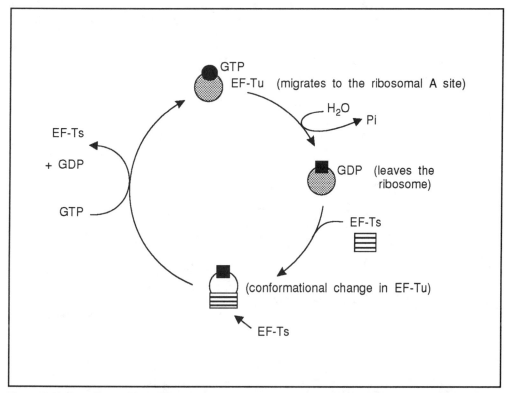

Figure 6.14 Recycling of EF-Tu-GTP. On delivering the amino-acyl tRNA to the A site of a ribosome, the GTP associated with EF-Tu is hydrolysed. This causes EF-Tu to leave the ribosome. Associating with a further elongation factor, EF-Ts, causes a conformational change such that bound GDP can be replaced by GTP. The re-primed EF-Tu is now available to participate further in protein synthesis.

6.5.3 Termination

release factors

As we saw in Section 6.2, the genetic code specified three stop codons or termination codons - UAA, UGA and UAG. When one of these enters the ribosomal A site, no amino-acyl tRNA binds. Instead two additional proteins, called release factors (RF1 and RF2), recognise the codons. RF1 recognises UAA and UAG, whereas RF2 is able to recognise UAA or UGA. On binding to a stop codon, these factors cause peptidyl transferase to hydrolyse the ester bond between the carboxyl terminus of the growing peptide chain and the tRNA to which it is bound ie RF1 and RF2 are able to alter the specificity of the enzyme. Subsequently, mRNA and uncharged tRNA leave the ribosome, which in turn dissociates into small and large subunits. IF3 maintains the small subunits as separate entities available for use in further rounds of translation of other mRNAs.

SAQ 6.4

List the important accessory proteins required for the three stages of translation and summarise their functions and any important features.

polysomes

One last point about prokaryotic translation is that mRNAs are at any given time being translated by many ribosomes. Each ribosome is associated with about 80 nucleotides of mRNA, and they all move in the same direction (5' → 3') so that several proteins are being produced at the same time (see Figure 6.15). These arrangements are known as polysomes and are clearly visible on electron micrographs.

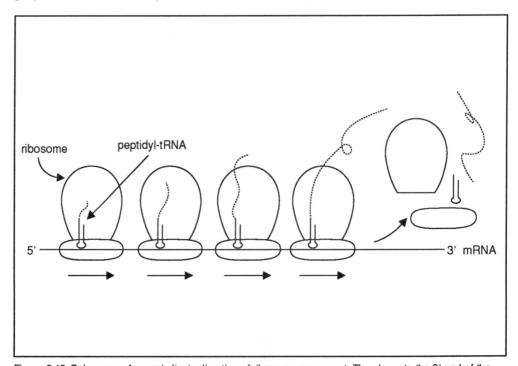

Figure 6.15 Polysomes. Arrows indicate direction of ribosome movement. The closer to the 3' end of the mRNA, the longer the nascent (growing, but incomplete) polypeptide chain. This is released when a termination codon is reached.

6.6 Eukaryotic translation

Although eukaryotes are not the subject of this volume it is worth mentioning that eukaryotic translation occurs by a very similar mechanism to that described above. The major differences are that:

- translation (in the cytosol) is physically, and hence temporaly, separated from transcription (in the nucleus);

- additional accessory protein factors are required for the initiation of eukaryotic translation;

- eukaryotes use a single release factor;

- the initiation codon for eukaryotic translation is always AUG, which codes for methionine rather than N-formyl methionine;

- eukaryotic ribosomes are larger than their bacterial counterparts and have an additional unique 5.8S rRNA.

Further details on translation in eukaryotes are given in the BIOTOL text 'Genome Mangement in Eukaryotes.'

Summary and objectives

In this chapter we have seen how the three major RNA species of *E.coli*, mRNA, tRNA and rRNA, participate in protein synthesis. mRNA provides the code (Section 6.2) for linking specified amino acids in the correct order into functional polypeptides. These amino acids are brought together along the mRNA base triplet codons by tRNA molecules (Section 6.3). The tRNA molecules thus act as adaptors and have a structure suited to this role. The anticodon (which recognises the mRNA codon by basepairing) is situated at one end of the molecule, with the amino acid attached to the 3′ terminus at the opposite end. The amino acids in themselves have no influence over which positions they are inserted into polypeptide chains. This is governed solely by codon: anticodon recognition. Hence the enzymes (amino-acyl tRNA synthetases) which produce amino-acyl tRNAs must be highly specific. The wobble hypothesis explains why each amino acid may only need one or two tRNAs to recognise all codons which specify it.

We have described how many protein factors are involved in the initiation of protein synthesis, chain elongation and termination of translation. We have also learnt that protein synthesis is energy demanding and results in the hydrolysis of ATP and GTP.

Now that you have completed the chapter you should be able to:

• describe experiments that enabled the genetic code to be elucidated;

• explain the wobble hypothesis and the advantages that arise from the genetic code being degenerate;

• describe the processes of initiation, elongation and termination as applied to translation;

• list the accessory proteins involved in translation and describe their roles.

Regulation of gene expression: control by repressor/activator proteins

Regulation of gene expression: control by repressor/activator proteins

7.1 Introduction

In previous chapters you have learnt about DNA structure and how this provides the genetic material which is transcribed into different species of RNA involved in protein synthesis. In this and the following chapter some of the ways in which prokaryotic gene expression is regulated will be discussed.

In this chapter we will consider the need for controlling gene expression, the ways in which genes are grouped together to simplify this control, and two specific examples - the *lac* and *ara* operons.

The term 'gene expression' covers the steps involved in the formation of a mature end product directed by a gene. Depending on the gene, this may be a protein, a tRNA molecule or an rRNA molecule.

∏   At which of the stages involved in production of a protein do you think gene expression can be controlled?

The first things to consider are the steps involved. These are transcription, translation and post-translational modifications (if appropriate). Control can occur by regulation of any, or all, of these stages, and examples of all types of regulation are known.

∏   Control over one of these stages is generally preferred, which do you think it is?

The answer should be fairly obvious, it is transcription - the first stage. It is not worth the cell expending energy on transcription if the encoded protein is not required. Hence it is control of transcription that we will consider here.

Control in prokaryotes is facilitated by the fact that genes encoding proteins with related functions are often grouped together and controlled as single units. Some of the ways in which control is exercised include the use of repressor (Section 7.3) and/or specific activator proteins (Sections 7.4 and 7.6) to modulate access of RNA polymerase to the structural genes. Specific examples of these will be considered. Before you continue on to this and the following chapter, it is important that you should be familiar with the principles of transcription and translation, so spend some time ensuring that you have mastered Chapters 5 and 6, and that you understand the distinction between transcription (RNA production) and translation (protein synthesis).

∏   Why should a simple organism such as *E.coli* need to regulate expression of its genes?

Even the simplest of bacterial organisms contains many, many genes to enable it to grow and divide and to cope with a potentially changing external environment. The products of these genes are not all required continuously, and indeed some may only be required every few generations. Hence to conserve metabolic energy, if their products are not required, genes are switched off. We simplify this concept by saying that there are two classes of genes: the 'housekeeping genes' which are more or less continuously expressed and the 'special function genes' which are only activated under specific conditions.

housekeeping genes

special function genes

∏ Write down some examples of housekeeping and special function genes.

Well, some obvious housekeeping genes are those coding for enzymes involved in everyday metabolic pathways such as glycolysis. Special function genes include those coding for enzymes involved in the catabolism (metabolic breakdown) of unusual substrates (eg lactose) under stress conditions. Other examples include genes coding for various anabolic enzymes (ie those involved in synthetic metabolic pathways). If the end product of the pathway directed by these enzymes can be obtained from the surrounding medium, it is a waste of energy for the bacterium to make it, and hence the genes involved are switched off.

In this chapter we will consider how two clusters of genes responsible for catabolism of occasional substrates are controlled. Additionally, the distinction between positive and negative control of gene expression, and examples of both, will be discussed, as will the concept of specific and of generalised controlling agents.

7.2 Bacterial promoter sequences

As you have seen in Chapter 5, initiation of transcription involves recognition of, and binding to, promoter sequences by RNA polymerase. Mutation experiments have indicated that two highly conserved regions within each promoter are necessary for RNA polymerase binding. These sequences are centred around base positions -10 and -35 relative to the start-point of transcription (designated +1) as illustrated in Figure 7.1. The distance between the two binding regions is approximately two complete turns of the double helix - ie both sites are on the same side of the DNA molecule. This may be important in allowing contact between the enzyme and its DNA template. Many promoters have slight differences from these consensus sequences, and the greater the difference the weaker the promoter. A weak promoter is one to which RNA polymerase binds more loosely or from which it initiates transcription less frequently. Thus, one way in which gene expression may be limited is by having a weak promoter associated with the coding region.

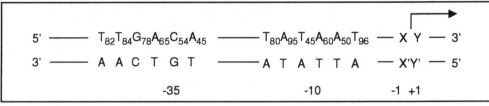

Figure 7.1 Bacterial promoter. The lines represent double-stranded DNA, with the -10 and -35 consensus sequences given in detail. The subscript numbers indicate the % of known bacterial promoters containing the given base at each position. To illustrate the numbering system involved, Y is the first base transcribed, the direction of transcription being indicated by the arrow. This position is numbered +1, and the preceding one (X), -1.

The RNA polymerase, once bound to the promoter, covers a large region of the DNA, extending across the start-point of transcription. This is illustrated for the operon in Figure 7.2. We will return to this when we consider control of the operon in Section 7.4.

The precise sequence between and around the -10 and -35 sequences appears to be relatively unimportant in terms of influence on RNA polymerase binding. However the overall context (as defined by the % base composition) may have an influence on promoter efficiency.

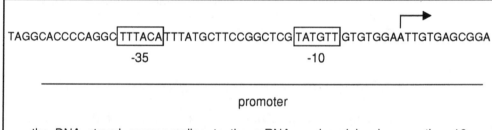

Figure 7.2 Promoter region of the *lac* operon.

7.3 Clustering of bacterial genes and principles of control

As alluded to above, bacteria and virus often contain clusters of genes within their genomes, which facilitates control. We have already mentioned some of these clusters, which are termed operons eg the *lac* and *trp* operons of *E.coli* - these encode proteins involved in lactose catabolism and tryptophan synthesis, respectively.

operons contain control elements and structural genes

Operons contain both control elements and structural genes as illustrated in Figure 7.3. The structural genes code for proteins involved in related processes, eg enzymes catalysing different steps in a common biosynthetic or catabolic pathway etc. The control elements occur upstream of (5' in relation to the coding strand), the structural genes and determine whether these genes will be transcribed or not. The genes are transcribed together to produce a single mRNA molecule called a polycistronic RNA. This mRNA is sequentially translated by ribosomes, giving rise to distinct polypeptides, one from each structural gene. This makes sense as either all or none of the enzymes/proteins will be required in the cell and clustering of the genes means that all the activities are co-ordinately controlled. The control region (X in Figure 7.3) contains promoter and operator sequences.

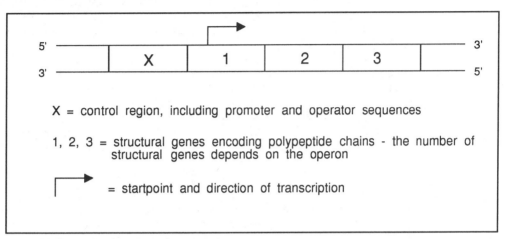

Figure 7.3 Generalised structure of an operon.

promoter
operator

Control over expression of the structural genes in many operons is exercised by regulation of the binding of RNA polymerase to the promoter. Control can be exercised by binding of repressor or activator proteins to the operator or other sites. 'Operators' are DNA sequences that bind proteins which often prevent transcription. Effectively, the bound protein can act as a physical barrier preventing RNA polymerase from interacting with the promoter and/or moving to the transcriptional start-point for expression of the structural genes within an operon. Other proteins may bind to operator sequences to enhance transcription.

repressor
activator

negative
control

Some operons are under negative control in that they are switched on unless a 'repressor' protein binds to the operator. Usually the repressor is expressed constitutively, and its ability to interact with the operator is governed by its interaction with a specific small molecule. The nature of this small molecule and its effect on the repressor protein reflect the function of the gene products as we shall see in the examples to be discussed later. The small molecule modulators act as co-repressors or as inducers. Co-repressors bind to inactive repressor proteins to activate them ie they allow the repressor to interact with the operon control region to block transcription. Inducers, on the other hand, prevent active repressors to which they bind from interacting with operators and hence activate gene expression. Binding of inducers and co-repressors to repressor proteins is readily reversible.

co-repressor
inducer

∏ See if you can identify what type of metabolic operons (coding for anabolic or catabolic enzymes) would benefit from inducible control and which from repressible negative control.

Let us think about inducible control first. Effectively, this situation involves the operon in question being switched off until the inducer is present. This is an ideal way to regulate a catabolic pathway. Such a pathway only needs to occur if the substrate is present. Hence the substrate (or some simple metabolite of it) may act as an inducer to activate its own metabolism. This is the case for the *lac* operon to be discussed in Section 7.4. On the other hand, if we think about an anabolic pathway - this only needs to be active whilst levels of its end product are low ie it needs to be switched on only until sufficient end product is present. This would be an ideal candidate for co-repressor control, with the metabolic end product acting as co-repressor. An operon controlled in this way is that encoding enzymes involved in tryptophan synthesis, (the *trp* operon), and this is discussed in Chapter 8. Other control mechanisms are also involved in the regulation of the *lac* and *trp* operons, and these are discussed at the appropriate places in this and the next Chapter.

positive control

In contrast to these forms of negative control, some operons are under positive control. In these cases, expression can only occur when an active 'activator' protein is present. This protein interacts directly with the control elements of the operon and/or the RNA polymerase, and may be induced or repressed by interactions with other small molecules in similar ways to those discussed for repressor proteins. An operon regulated by both positive and negative control is the *ara* operon, discussed in Sections 7.6 and 7.7.

SAQ 7.1

Draw four brief sketches to indicate how positive and negative control of prokaryotic gene expression can occur, showing how small inducer and repressor molecules may influence the proteins involved, ie distinguish between inducible negative and positive regulation and repressible negative and positive regulation of gene expression.

7.4 The *lac* operon of *E. coli*, structure and control by lactose level

promoter, operator, 3 structural genes

This operon is illustrated in Figure 7.4. It codes for proteins involved in the transport and breakdown of lactose and contains two control elements (promoter and operator) and three structural genes. The latter code for the enzymes β-galactosidase (coded by *lacZ*), galactoside permease (*lacY*) and galactoside transacetylase (*lacA*). β-galactosidase catalyses the breakdown of lactose to glucose and galactose and galactoside permease allows entry of lactose into the bacteria, although the function of the third enzyme is uncertain. Galactoside transacetylase catalyses the transfer of acetyl groups onto β-galactosides, and it is though that this may be important for modifying molecules which are not β-galactosidase substrates. The modified molecules may then be otherwise degraded or excreted by the cell.

∏ Write down a reason why this action may be important to the cell?

One explanation may be that the galactosides in question are capable of entering the cell via the permease, and of acting as inducers of the operon. Unless they are removed, they would cause the operon to be in an activated state even if no lactose was present. This would cause the cell to waste energy on unnecessary transcription and translation.

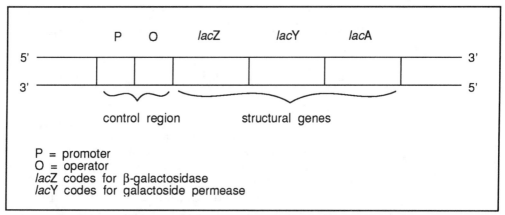

Figure 7.4 Overview of the *lac* operon. The promoter and operator regions of the operon exert control over transcription of the structural genes. They accomplish this by binding RNA polymerase and a repressor protein respectively. For transcription to occur, RNA polymerase binds to P and moves through O on to the structural genes. An active repressor protein bound to O forms a physical barrier, preventing RNA polymerase access to the structural genes.

lacI gene   The operon is under negative control by a repressor protein coded by the *lacI* gene. This is situated just upstream of the *lac* operon, but is not part of the operon. The *lacI* gene is continuously expressed at a low level, so that repressor protein is always present. This protein binds to the operator sequence of the operon, preventing transcription. The repressor was thought to prevent polymerase binding because the promoter and operator regions overlap. This is illustrated in Figure 7.5, which is a more detailed picture of Figure 7.2. As shown in Figure 7.5b, the operator region is interesting as it has a twofold axis of symmetry (indicated by the arrows). This is not perfect, but involves 16 out of 21 positions.

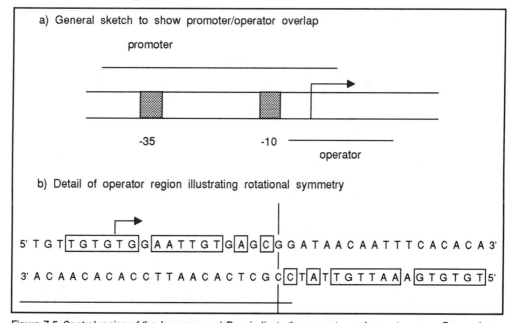

Figure 7.5 Control region of the *lac* operon. a) Bars indicate the promoter and operator areas. Boxes show the -10 and -35 sequences. b) Boxed regions indicate identical sequences in the two halves of the operator. Bar shows region which overlaps the promoter. Vertical line indicates the axis of the rotational symmetry.

∏ What might this repeated sequence within the operator indicate about interaction with the repressor?

The symmetry may indicate that the repressor protein exists as a dimer of identical subunits each of which contacts the operator. Indeed, X-ray crystallography studies have confirmed this to be the case for a variety of repressor proteins.

trans-acting element

The repressor protein is known as a 'trans-acting' element (or factor) in that it is coded for by a gene outside the operon, although exerting influence over it. Indeed, the *lacI* gene could be located anywhere in the genome and still exert its influence on the operon. The operator and promoter regions, on the other hand, are called 'cis-acting' elements or sequences. These elements must be associated directly with the structural genes that they control, and it is the DNA itself that performs the regulatory function.

cis-acting elements

inducible negative control

In the absence of lactose, the repressor protein binds to the operator, excluding access of RNA polymerase (which binds initially to the promoter) to the structural genes and preventing transcription of *lacZ*, *lacY* and *lacA*. However, each bacterium retains between three and five molecules of each of the gene products. This is important, as without any permease, no lactose could enter the cell even if it became available in the external medium. Also, a low level of β-galactosidase is required, as it converts a small proportion of the lactose entering the cell into a form that acts as the operon inducer. The major reaction catalysed by β-galactosidase is the hydrolysis of lactose to glucose and galactose. However, the enzyme also converts a small proportion of lactose into allolactose (see Figure 7.6). Allolactose binds to the repressor protein causing it to change its conformation such that it dissociates from the operator. RNA polymerase may now initiate transcription. The allolactose-repressor interaction is reversible, and when the level of lactose, and hence allolactose, in the cell decreases the repressor is freed for further interaction with the operator. This type of control over gene expression is an example of inducible negative control.

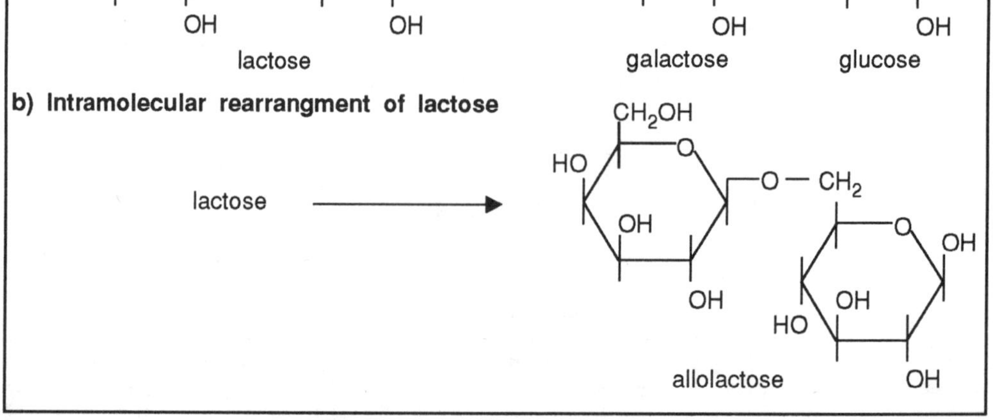

Figure 7.6 Actions of β-galactosidase.

Recent research has shown that although the overall regions of DNA covered by the repressor and RNA polymerase overlap (Figure 7.5), the proteins actually specifically interact with directly adjacent sites on the DNA. Hence, RNA polymerase is bound at the promoter even when the repressor interacts with the operator. Repressor binding has been shown to enhance RNA polymerase binding to the promoter. This is useful as it means that the instant allolactose liberates the repressor, RNA polymerase can start transcription. This relationship may not hold true for all operons.

SAQ 7.2

Certain mutant *E.coli* strains exist which have been useful in determining the precise mode of control of the *lac* operon. Some of these strains are described below:

strain A	genotype	$lacI^+ \ lacZ^+ \ lacY^+ \ lacA^+$
strain B		$lacI^- \ lacZ^+ \ lacY^+ \ lacA^+ \ O^+ \ P^+$
strain C		$lacI^+ \ lacZ^+ \ lacY^+ \ lacA^+ \ O^- \ P^+$
strain D		$lacI^+ \ lacZ^+ \ lacY^+ \ lacA^+ \ O^+ \ P^-$
strain E		$lacI^+ \ lacZ^+ \ lacY^- \ lacA^+ \ O^+ \ P^+$
strain F		$lacI^+ \ lacZ^- \ lacY^+ \ lacA^+ \ O^+ \ P^+$
strain G		$lacI^+ \ lacZ^+ \ lacY^+ \ lacA^- \ O^+ \ P^+$

+ indicates gene element is intact as in wild type *E. coli*, strain A

- indicates gene element is non-functional

Using the information you have obtained so far in this chapter, and remembering that prokaryotes are normally haploid (ie contain only one copy of each chromosome), explain the state of expression of the *lac* operon in the presence and absence of lactose in these strains.

SAQ 7.3

E.coli may be made partially diploid experimentally, by introduction of exogenous DNA. Consider what would happen to the following strains (named as in SAQ 7.2) if pieces of DNA containing the genes/control elements shown are introduced, making them partially diploid. Explain what the state of operon expression will be in the strains in the presence and absence of lactose:

1) Strain B + $lacI^+$

2) Strain C + O^+

3) Strain D + P^+

4) Strain E + $lacY^+$

5) Strain F + $lacZ^+$

6) Strain G + $lacA^+$

7) Strain A + $lacI^-$

NB remember, the introduced DNA will be separate to the genomic DNA, not integrating into it, ie a separate strand.

7.5 The *lac* operon control by glucose level

The impression you might have gained so far is that the operon is either switched fully on or fully off, depending on the presence or absence of the specific inducer, allolactose. This is not, in fact, the case, as this operon is one of many influenced additionally by the availability of glucose as a metabolic substrate. *E.coli* preferentially uses glucose as a fuel if given the choice, even if other possible substrates such as lactose are present. The reason for this is unclear, but may be a reflection of the evolution of the organism.

cAMP

So, the next question is how is the *lac* operon influenced by the available glucose level? Basically, the level of glucose influences the cellular level of cAMP (5′3′ cyclic AMP: Figure 7.7), which in turn regulates operon activity. When glucose levels are high, the amount of cAMP present is reduced and vice versa ie an inverse relationship exists. It is though that the enzyme, adenylate cyclase, responsible for formation of cAMP is regulated by a glucose metabolite. The cAMP produced exerts a positive control over operon expression by combining with the activating protein called either CAP (catabolite activator protein) or CRP (cAMP receptor protein) which enhances RNA polymerase activity. Thus when glucose levels are high, cAMP levels are low, CRP is not activated and RNA polymerase activity is not enhanced.

CAP or CRP
protein

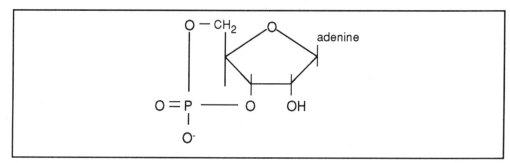

Figure 7.7 Cyclic adenosine monophosphate (cAMP).

positive control

catabolite
repression

Conversely, when not much glucose is available and the cell needs to make maximum use of whatever other metabolic substrates there are, cAMP levels rise, CRP is activated and the rates of transcription of various appropriate operons are enhanced. Hence control by cAMP/CRP is an example of positive control of bacterial gene expression, albeit a non-specific signal influencing various unrelated operons. This phenomenon of genetic control by glucose is known as catabolite repression - high levels of glucose (the preferred catabolite) inhibit other catabolic operons.

CRP itself is a dimeric protein of identical polypeptide chains containing 210 amino acids. Binding of cAMP causes it to undergo a conformational change such that it can interact with DNA. The site within the operon where it binds (the CRP site) is shown in Figure 7.8, covering positions -52 to -72 ie upstream of the promoter site but downstream of the *lacI* gene. On binding to this site, cAMP/CRP complex enhances RNA polymerase initiation from the promoter, increasing the overall rate of transcription. In the absence of the cAMP/CRP complex, little transcription occurs of the operon, even in the presence of lactose. The CRP binding sites of different operons vary somewhat. The consensus sequence of this region is shown below, the two boxed regions being almost inverted repeats of one another, and being the most highly conserved regions:

5' | AANTGTGA | NNTNNN | TCANATT | 3'

3' | TTNACACT | NNANNN | AGTNTAA | 5'

where N is any nucleotide.

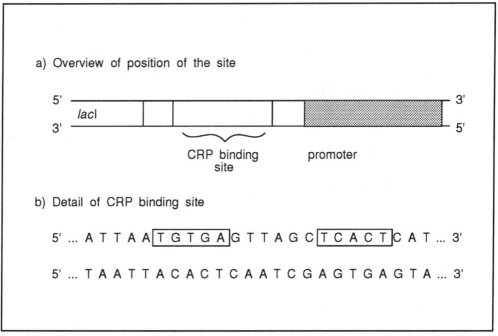

Figure 7.8 CRP binding site of the *lac* operon. Boxes indicate the most important positions in the binding site. These regions are fairly well conserved between different operons, and mutations in them alter cAMP/CRP binding ability.

Precisely how CRP exerts its effects on transcriptional initiation is unclear, as the CRP site occurs at different positions within various operons relative to the promoter. As we have seen, in the *lac* operon this site is just upstream of the promoter, in the *gal* operon it occurs within the promoter (indicating a very close contract between CRP and RNA polymerase), and in the *ara* operon, the CRP site is a fair distance upstream of the promoter (-107 to -78). In the latter case, another protein is known to bind between CRP and RNA polymerase and this is discussed in more detail in Section 7.7. One possibility that has been suggested for the mode of action of CRP is that it helps correct interaction of the RNA polymerase with the DNA prior to melting (or separation of the two strands) of the template. This suggestion arose due to the finding that most CRP-responsive operons have -35 sequences which deviate markedly from the consensus sequence (Figure 7.1). This region is important in transcriptional initiation, and it may be that CRP binding substitutes for polymerase/-35 interaction, and subsequently allows template melting. The precise mode of action is, however, still controversial and awaits further research for its determination.

Recently a secondary role of CRP as a repressor in various operons has been suggested. In addition to the major promoter of the *lac* operon described in detail above, there is another upstream promoter. This overlaps the CRP site, and binds RNA polymerase with high affinity. However, once bound the polymerase does not initiate transcription efficiently from this promoter. cAMP/CRP binding to the CRP site prevents polymerase binding to this weak (in terms of gene expression) promoter, and enhances binding to the strong major promoter. Hence cAMP/CRP acts as a repressor of the weak promoter.

∏   Before continuing, summarise the way in which glucose influences the *lac* operon expression, as we will return to this concept later.

non-specific and operon-specific control

Basically, high glucose levels repress the operon by reducing the level of cAMP and hence of active cAMP/CRP complex. This complex is required to allow efficient initiation by RNA polymerase of operon expression. This form of control is non-specific in that CRP influences several operons, unlike the effect of allolactose which is operon-specific. Additionally, CRP represses transcription from a weak secondary promoter within the operon, effectively directing RNA polymerase to the stronger one. cAMP/CRP-activation is an example of positive control of gene expression. Transcription cannot occur at a significant level unless the complex is bound to the operon. A more specific example of positive control is that of the *E. coli* arabinose operon (*ara* operon) described in the next section.

7.6 The *ara* operon: general features

This operon, also called *ara* BAD, encodes the enzymes responsible for conversion of arabinose to xylulose-5-phosphate, a metabolite which can enter the pentose phosphate pathway. Arabinose is an aldopentose sugar which is an important constituent of both plant and bacterial cell walls. The pathway for its conversion to a useful metabolic intermediate, the enzymes involved, and an overview of the operon are shown in Figure 7.9. The *ara* A gene product, arabinose isomerase, converts the sugar to its ketose isomer, ribulose. This is activated by phosphorylation (catalysed by the *ara* B gene product, ribulokinase) and converted to the C4 epimere, xylulose 5-phosphate by the *ara* D product, ribulose 5-phosphate epimerase. *ara* C encodes a positive regulator of the operon, C protein. To activate transcription of *ara* BAD, C protein must first bind arabinose as an activator. In other words, the operon is only active when the substrate of the encoded enzymes is present. In addition to the specific regulator, C protein, the operon is responsive to cAMP/CRP, ie it is controlled by the influence of two activator proteins - one specific and a more general one which is regulated by the glucose level. As you might expect, this means that the overall control mechanism is quite complex. We will look at this in stages to try to avoid confusion.

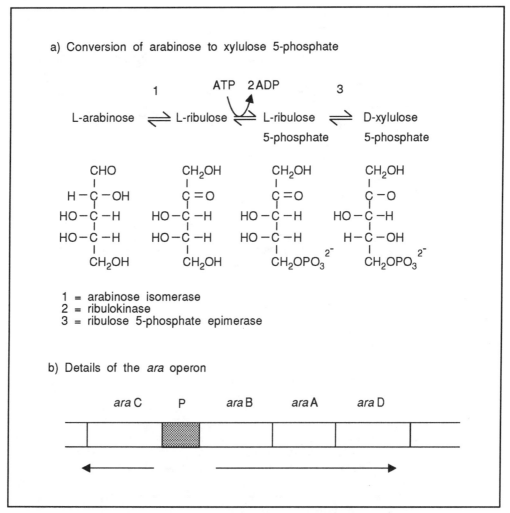

a) Conversion of arabinose to xylulose 5-phosphate

1 = arabinose isomerase
2 = ribulokinase
3 = ribulose 5-phosphate epimerase

b) Details of the *ara* operon

Figure 7.9 The *ara* operon. *ara* C encodes C protein which exerts control over the structural genes. P represents the control region, which is more complex than that of the *lac* operon (see text). *ara* B encodes the kinase, *ara* A encodes the isomerase and *ara* D encodes the epimerase. Arrows indicate the direction of transcription.

7.7 The *ara* operon: the control region

In addition to activating operon transcription, the C protein also represses it under certain conditions, and can do this in two ways. To understand these actions we need firstly to look at the *ara* operon control region in a little more detail. This is illustrated in Figure 7.10 and contains two promoters: one for the *ara* C gene and the other for *ara* BAD. In addition there are three C protein binding sites and a CRP site. So, how do interactions of CRP and C protein with these sites regulate *ara* operon expression? To start with, let us consider the C protein binding site associated with the *ara* C promoter. This site binds C protein when it reaches a certain level in the cell.

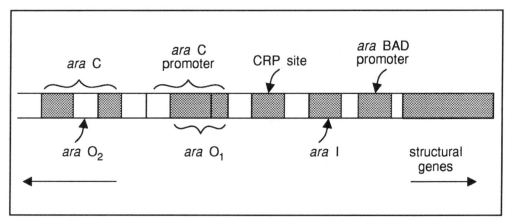

Figure 7.10 Detail of the *ara* operon control region. Arrows indicate direction of transcription. CRP site = cAMP/CRP binding site. *ara* O_2, *ara* O_1 and *ara* I are C protein binding sites. *ara* O_2 is located within the *ara* C gene, and *ara* O_1 overlaps the promoter of this gene. See text for further details.

⊓ What will the result of C protein/*ara* O_1 binding be?

It should be apparent that binding of C protein to *ara* O_1 will prevent RNA polymerase from attaching to the *ara* C promoter, preventing production of further C protein. This acts as a form of negative feedback or autoregulation - once sufficient C protein is available its synthesis is switched off, and levels thus remain fairly constant in the cell.

The other two C protein binding sites are responsible, together with the CRP binding site, for control of *ara* BAD expression. Firstly, let us consider what happens in the presence of arabinose and absence of cAMP/CRP (ie when glucose levels are high). Under these conditions a molecule of C protein binds to both *ara* O_2 and *ara* I, producing a loop in the DNA as shown in Figure 7.11. This is a repressed form of the operon and C protein cannot activate RNA polymerase to transcribe from the *ara* BAD promoter even though arabinose is present.

⊓ The *ara* O_2 and *ara* I sites are approximately 200bp apart. What does this suggest about their interaction with C protein?

Each complete turn of the DNA double helix occurs every 10bp, so these two binding sites are on the same side of the helix. This is necessary if C protein is to contact them both in the looped structure described. Indeed, insertion or deletion of sequences in the intervening region such that the spacing is of an incomplete number of helical turns abolishes loop formation.

So, what happens if cAMP/CRP is present in addition to arabinose? In this case, the CRP site next to *ara* I (see Figure 7.10) is occupied, changing the action of C protein. It releases *ara* O_2, but remains bound to *ara* I allowing activation of RNA polymerase at the *ara* BAD promoter, and transcription of the structural genes. It is uncertain how CRP exerts its effect on C protein, but it may be by direct contract between the molecules as they bind to adjacent sites on the DNA.

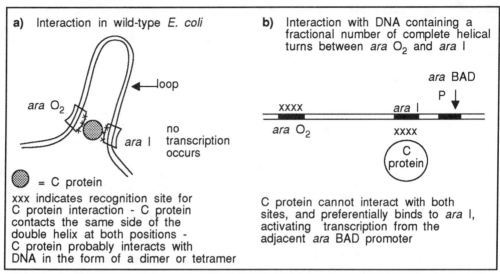

a) Interaction in wild-type *E. coli*

←loop

ara O$_2$

ara I   no
transcription
occurs

= C protein

xxx indicates recognition site for
C protein interaction - C protein
contacts the same side of the
double helix at both positions -
C protein probably interacts with
DNA in the form of a dimer or tetramer

b) Interaction with DNA containing a
fractional number of complete helical
turns between *ara* O$_2$ and *ara* I

ara BAD
P ↓

xxxx   *ara* I

ara O$_2$ xxxx

C
protein

C protein cannot interact with both
sites, and preferentially binds to *ara* I,
activating transcription from the
adjacent *ara* BAD promoter

Figure 7.11 Interactions of activated C protein with *ara* O$_2$ and *ara* I in the absence of cAMP/CRP.

∏ Bearing in mind what we have said about the actions of C protein so far, what do
you think would be the effect on control of operon expression of deletion of the
ara O$_2$ site (assuming functional C protein was still made)?

Well, if no *ara* O$_2$ site is available, no repressor loop will form either in the presence or
absence of cAMP/CRP. Hence, as long as C protein and arabinose are present, the
operon will be active and not influenced by the glucose level.

SAQ 7.4

Draw a table to compare the two operons you have learnt about in this chapter.
Some of the parameters you may like to compare are:

1) Inducer/activator.

2) Operator(s) involved.

3) Promoter(s) involved.

4) Effect of absence of inducer/activator.

5) Effect of addition of inducer in the presence of glucose.

6) Effect of addition of inducer in the absence of glucose.

7.8 Final comments

As might be expected, control of eukaryotic gene expression follows a much more
complex pattern, but in fact does show some similarities to that described above.
Generally, eukaryotic expression at the level of transcriptional initiation is also formed
by protein-promoter interactions. Eukaryotic RNA polymerases, however, are unable
to interact directly with gene promoters (unlike their prokaryotic counterpart) and
show an absolute requirement for prior binding of ancillary transcription factors to
specific promoter regions. The more complex topic of control of eukaryotic gene
expression is the subject of the BIOTOL text 'Genome Management in Eukaryotes'.

Summary and objectives

In this chapter we explained that prokaryotes simplify the control of gene expression by grouping together genes of related function into operons. We illustrated this by in depth discussion of the *lac* operon in *Escherichia coli*. We discussed the strucutre of bacterial promoter and operator sequences and explained how the *lac* operon is under negative control by a repressor protein and described the positive regulation of this operon by CRP. The control of this operon through the effects of glucose levels on adenylate cyclase in the production of cAMP and its effect on CRP were also described.

We also described the regulation of the *ara* operon. This operon is under the positive control of C protein, a product of an *ara* gene. The operon is also regulated by cAMP/CRP.

Now that you have completed this chapter you should be able to:

* give examples of housekeeping and special function genes;

* describe the promoter region of the *lac* operon in terms of -10 and -35 sequences;

* describe the structure of the *lac* operon in *E. coli* with details of the promoter region of this operon;

* explain how glucose can influence the expression of the *lac* genes through the mediation of adenylate cyclase and CRP;

* identify cis- and trans-acting genes;

* describe the principal features of the *ara* operon and explain the roles of the C protein and cAMP/CRP in the regulation of this operon;

* make a comparison between the regulation of the *lac* and *ara* operons.

Control of gene expression: the *trp* operon and other regulatory mechanisms

Control of gene expression: the *trp* operon and other regulatory mechanisms

8.1 Introduction

In the previous chapter we have seen how some operons are controlled at the level of initiation of transcription by the presence of particular proteins and associated modulator molecules. This, however, is not the only way in which expression can be controlled at the transcriptional level. Other operons are regulated after initiation of transcription by premature termination of the transcript. This occurs due to the formation of specific secondary structures within the transcript, and is known as attenuation.

In Chapter 7, we considered two relatively simple examples of prokaryotic genetic control - those adopted by the *lac* and *ara* operons. However, many operons are under the influence of more than one specific control mechanism. The example which we shall consider in detail in this chapter is the *trp* operon of *E.coli*. This operon codes for enzymes involved in the biosynthesis of tryptophan from chorismate as shown in Figure 8.1. Anthranilate synthetase is composed of two different 60kd subunits, the products of two genes, *trp*E and *trp*D. Indole glycerol-phosphate synthetase is a 45kd protein, the product of the *trp*C gene. Tryptophan synthetase is an $\alpha_2\beta_2$ tetramer, the α subunits being coded by *trp*B, and the β subunits by *trp*A. We shall look at the arrangement of these genes in the *trp* operon in Section 8.2. Control over the genes involves both repression (ie control of transcriptional initiation) and control over transcriptional termination, termed attenuation. It would be a good idea to make a table listing the various genes (eg *trp*A, *trp*B) and their products.

Attenuation basically involves premature termination of transcription before the structural genes of an operon are reached.

Π   What do you think is most likely to trigger attenuation of *trp* operon expression?

The answer is obvious really, it is tryptophan itself in the form of tRNA^{trp} (that is tryptophan tRNA loaded with its amino acid). This operon encodes enzymes for tryptophan synthesis, so if tryptophan is abundant these are not required and the structural genes need not be transcribed. We shall consider how the response to the level of tryptophan occurs in Section 8.4. Additionally, other operons under control by attenuation will be considered to consolidate your understanding of this process in Section 8.5.

Towards the end of this chapter we shall consider a couple of other ways in which gene expression is regulated. In Section 8.6 we shall see how production of antisense RNA reduces effective expression, and how production of different σ subunits in response to different external stimuli is important to micro-organisms.

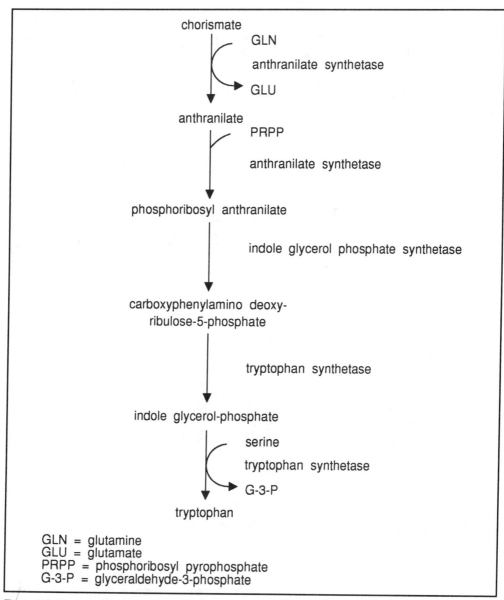

Figure 8.1 Biosynthetic pathway for the formation of tryptophan in *E.coli.*

8.2 The *trp* operon

8.2.1 Structure of the *trp* operon

leader
sequence

As mentioned above, this operon contains five contiguous structural genes in addition to an upstream regulatory region. An overview of *trp* operon structure is shown in Figure 8.2a. The difference between this control region and that of the *lac* operon is the presence of a 162bp 'leader region' between the operator and the first structural gene. This leader sequence forms the first part of any message transcribed by RNA polymerase, and contains within it a sequence ('a' in Figure 8.2a) which signals attenuation. Hence there are two possible transcription products of this operon shown in Figure 8.2b - a short 'attenuated' one solely containing the leader sequence and formed when tryptophan levels are high, and a full length one encompassing the structural genes, formed when tryptophan levels are low. We shall consider the role of the attenuator in Section 8.3, but we will start with the easier task of considering control of transcriptional initiation at the operon.

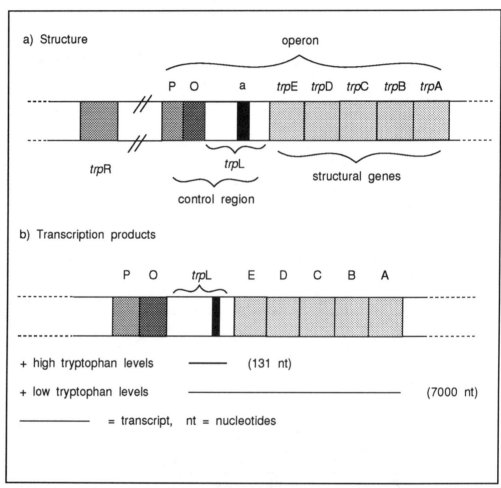

Figure 8.2 The *trp* operon. *trp*R is not actually part of the operon. It codes for the *trp* operon repressor. P = promoter, O= operator, *trp*L = operon leader sequence, this is 162 bp long, a = attenuator sequence. *trp*E and *trp*D encode subunits of anthranilate synthetase, *trp*C encodes indole glycerol phosphate synthetase, *trp*B and *trp*A encode subunits of tryptophan synthetase.

8.2.2 Control of transcriptional initiation at the *trp* operon

negative control   The *trp* operon is under negative control by a repressor protein (cf the *lac* operon), the product of *trp*R. *trp*R is constitutively expressed although its product is unable to interact with the operator unless it binds tryptophan (Figure 8.3).

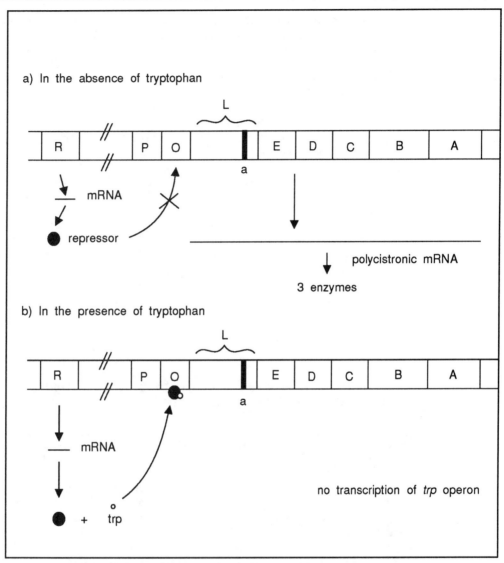

a) In the absence of tryptophan

b) In the presence of tryptophan

Figure 8.3 Repressor protein control of *trp* operon expression. a) Repressor alone cannot bind to operator, hence RNA polymerase can move from the promoter to transcribe the operon. b) In the presence of co-repressor, tryptophan, repressor binds to the operator to prevent transcriptional initiation.

As is the case with allolactose binding to the repressor, tryptophan-repressor binding is reversible, so that if tryptophan levels fall, the complex dissociates and the repressor leaves the operator. When bound to the operator, the repressor prevents RNA polymerase from initiating transcription.

∏ Would you expect the *trp* operon to be under positive control by CRP like the *lac* and *ara* operons?

No, of course not. The *trp* operon codes for anabolic enzymes. The presence or absence of glucose as a catabolic substrate need have no direct influence on expression of this operon. Alternatively, the *lac* and *ara* operons encode catabolic enzymes involved in the use of lactose and arabinose as energy sources. As *E.coli* prefers to use glucose, if available, control by cAMP is made use of in these cases.

It is important to stress at this point that, as in the case of the *lac* and *ara* operons, control by repression-induction is not absolute. By this we mean that an operon may not be completely activated or completely silenced under a given set of physiological conditions. A range of activities will exist, and a very low level of transcription is always likely to occur. The difference in level of expression in the presence and absence of activated repressor is 70-fold for the *trp* operon.

Early studies of the *trp* operon indicated that a second type of regulation was also involved in determination of the overall level of expression. The first indication of this was the observation that by varying the conditions of culture of *E.coli*, activities of the enzymes coded by the *trp* operon could be varied over a 600-fold range. This is far greater than the range alluded to above caused by repressor/activator protein-operator interaction, and shows an interplay between this type of interaction and, in this case, control by attenuation.

SAQ 8.1	Compare the ways in which repressor proteins controlling the *lac* and *trp* operons work.

8.3 Control of the *trp* operon by transcriptional termination - the basic principle of attenuation

∏ See if you can write down any possible reasons why a bacterial cell should use two regulatory mechanisms which both respond to the same signal (in this case the level of tryptophan)?

At first sight this does seem odd, however as the responses of the two systems will occur over different ranges of tryptophan concentration it actually lends more versatility to the system. This is shown by the large range of efficiency over which the operon may be expressed as indicated above. Generally, control by repression is exercised at higher tryptophan concentrations than control by attenuation. Additionally, attenuation is controlled by the level of tRNA^{trp} rather than by that of free tryptophan. We shall think about what this actually means a little later (see SAQ 8.2), after we have discussed attenuation.

attenuation

To begin this discussion, we must just recap on how transcription is terminated. Basically, there are two mechanisms - one is called 'rho-dependent' and the other 'rho-independent', referring to the requirement for an additional protein factor. Attenuation involves rho-independent termination. This occurs on recognition of a specific structure in the transcript by RNA polymerase.

∏ What are the features of the rho-independent terminator structure?

function of
hairpin loop
As you should recall, these are a hairpin loop followed by a run of approximately six U residues, as shown in Figure 8.4. Termination occurs close to the end of the run of Us. The hairpin structure in the transcript causes the enzyme to pause (for up to 60 sec) and the template-transcript hybrid has time to unwind. The basepairing between rU (in the transcript) and dA nucleotides (on the template) is the weakest possible combination, and breakage of these weak bonds at the terminator is favoured whilst the RNA polymerase pauses. Indeed, if the U residues are replaced by other nucleotides or their number is reduced, the enzyme continues transcription following the pause at the hairpin, and termination does not occur.

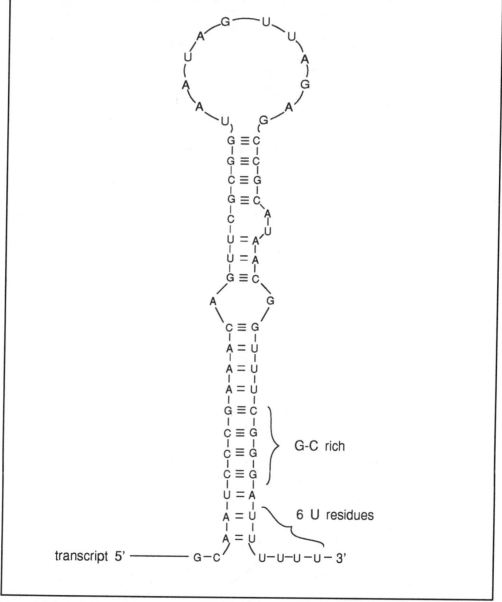

Figure 8.4 A rho-independent terminator. Hairpin loop formed due to self-complementary sequence in the transcript. The hairpin contains a G-C rich region and ends with a run of about 6 U residues. Adapted from B Lewin (1990), Genes IV, Oxford University Press, p 285, Figure 15.1.

The next thing to consider is the nature of the leader sequence. This sequence occurs between the operator and structural genes, and is essential for attenuation. As indicated above, the leader sequence is transcribed. In the presence of high levels of tryptophan only the first 140bp are transcribed (up to the attenuator), whereas in the absence of tryptophan the whole leader sequence is copied and transcription continues into the structural genes. The sequence of the attenuator is shown in detail in Figure 8.5 a). As indicated in the figure, it contains a GC rich sequence, which has a two-fold axis of symmetry, followed by an AT-rich region. The corresponding mRNA produced is shown in Figure 8.5b) and the rho-independent terminator structure that it can form in Figure 8.5c). As indicated, part of the leader upstream of the attenuator contains an open reading frame, encoding a 14 amino acid leader peptide. This sequence is close to a functional ribosome binding site which is capable of driving translation, although no free leader peptide is detected in normal cells. The leader peptide does not survive *in vivo* as it serves no purpose as a peptide per se. So why is it made? The answer to this is central to the mechanism of attenuation.

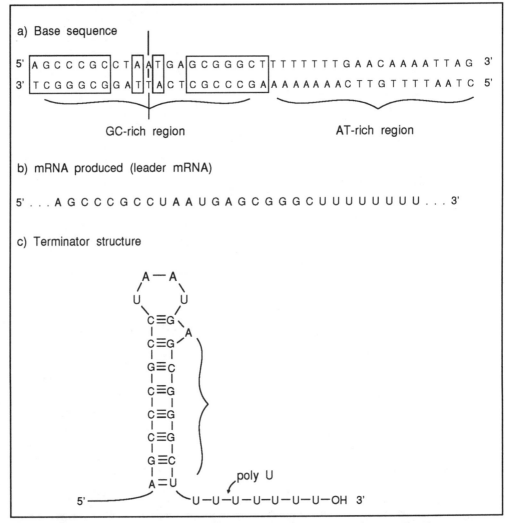

Figure 8.5 The *trp* operon attenuator. a) Base sequence: vertical line indicates the axis of symmetry. Boxes highlight symmetrical sequences. b) mRNA from the attenuator region. c) Terminator structure (see text for details).

∏ Look at the amino acid sequence of the leader peptide - what strikes you about this? (Figure 8.6 might help you).

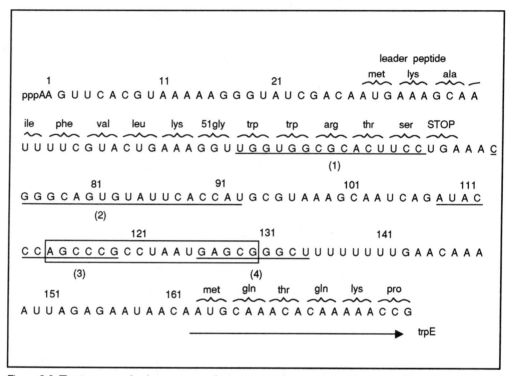

Figure 8.6 The *trp* operon leader sequence. the sequence given corresponds to the mRNA transcribed from the leader region. The numbers above the sequence refer to the nucleotide positions. Amino acid sequences coded within the leader and by the start of the first structural gene, *trp*E, are shown. The attenuator sequence is within the box. Sequences shown underlined are important in forming stem-loop structures, and are labelled as 1), 2), 3) and 4). The arrow indicates the position and direction of transcription of the *trp*E gene. See text for further discussion.

What should leap out at you is that the peptide contains two tryptophan residues, coded by adjacent codons. Thus, if this region is being translated ribosomes will only be able freely to negotiate it if there is sufficient tryptophanyl-tRNA. In the absence of tryptophan, the ribosomes will stall at this point. We will return to the importance of this in a little while.

stem-loop structures   Another important feature of the leader region is that it contains four stretches of sequence capable of forming various stem-loop structures. These sequences are indicated in Figure 8.6, and for ease of reference they have been numbered 1) - 4). Part of sequence 3) and sequence 4) basepair to one another to form the terminator structure shown in Figure 8.5c. Regions 1 and 2 may form a stem-loop structure as may regions 2 and 3.

∏ By drawing out the sequences of regions 1, 2 and 3, try to decide the optimal base-pairing for the production of the two stem-loop structures mentioned. Then examine Figure 8.7.

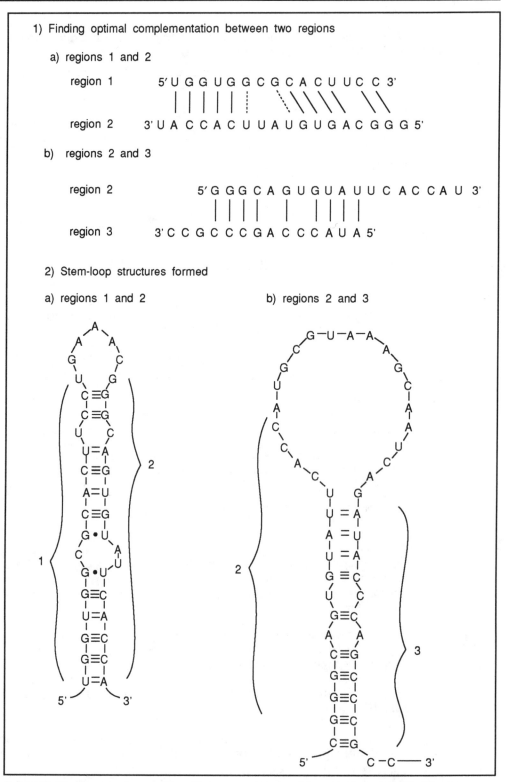

Figure 8.7 Stem-loop structures produced between regions 1 and 2, and between 2 and 3 of the *trp* operon leader region.

The optimal structures are shown in Figure 8.7 b). To determine the regions of maximum complementation it is easiest to write out each of the sequences on a separate piece of paper, writing one sequence $5' \rightarrow 3'$ and the other $3' \rightarrow 5'$. By moving the two pieces of paper with respect to one another a best fit can be achieved. (Remember, you can allow for several mismatches and for unequal lengths in the two strands between complementary bases). This is illustrated in Figure 8.7 a). Actually, this is the second best way of finding stem-loops. The best way is to use a computer!

Altogether then, we have three possible stem-loop structures which may form in a transcript of the leader region, some of which are mutually exclusive. For example, if the stem-loop between the regions 2 and 3 forms, region 3 will not be available to form
attenuator the terminator structure (also called the attenuator) with region 4. For this reason the region 2 - region 3 stem-loop is known as the anti-terminator or anti-attenuator
anti-attenuator structure. The anti-terminator itself can only form if region 2 is available and not already base-paired to region 1. Hence, overall control of transcription by attenuation relies on the stem-loop structures produced, which in turn are governed by translation of the leader peptide as we shall now see.

8.4 The influence of tryptophan levels on *trp* operon: transcriptional control by attenuation

For the time being, we shall assume that RNA polymerase has escaped regulation at the operator and has started to transcribe the leader region. The polymerase will transcribe through the region encoding the leader peptide and through stem-loop forming regions 1 and 2.

⫪ At this point the polymerase will pause - why is this?

If you remember, we learnt earlier that RNA polymerase always pauses after a stem-loop structure forms in the transcript. This is important for rho-independent termination if the stem-loop is followed by a polyU stretch. This is not the case with the structure formed here, so transcription does not terminate. However, the pause is vital as it allows a ribosome to interact with the transcript upstream of the leader peptide initiation codon. The ribosome rapidly starts to translate the leader peptide, and can disrupt the stem-loop structure to continue this action. Once this happens the polymerase is free to continue transcription past stem-loop region 2. The polymerase and ribosome thence move in unison, the enzyme just ahead of the ribosome.

We now need to consider what happens in the situations of abundance or absence of tryptophan. In the presence of tryptophan, there will be plenty of charged tRNA^{trp} to satisfy the requirement for formation of the leader peptide, and the ribosome will move through to the UGA stop codon. As we saw in Chapter 6, ribosomes are large structures and cover a significant region of RNA. When at the stop codon of the *trp* leader region, the ribosome will cover from approximately nucleotide 59 to 81 (Figure 8.6) ie a large part of the 5' end of stem-loop region 2.

⫪ What will this mean with respect to the stem-loop structures produced as RNA polymerase carries on along transcribing the leader region?

This will mean that the region 2 - region 3 stem-loop structure will not form completely. Only the four 5' most nucleotides of region 3 will basepair with nucleotides 82-85 in region 2. Hence the whole of the 3' end of region 3 necessary for formation of the attenuator/terminator will be available. On formation of this structure, transcription ceases and the structural genes (downstream of this region) are not expressed. This is illustrated in Figure 8.8.

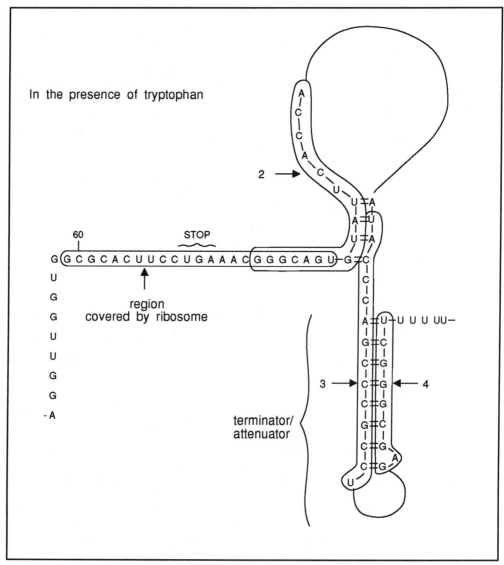

Figure 8.8 Structure of *trp* operon leader region transcripts in the presence of tryptophan. 5' end region 2 is unavailable for binding to region 3, which thus forms the terminator with region 4. Redrawn from B Lewin (1990), Genes, Oxford University Press, p 291.

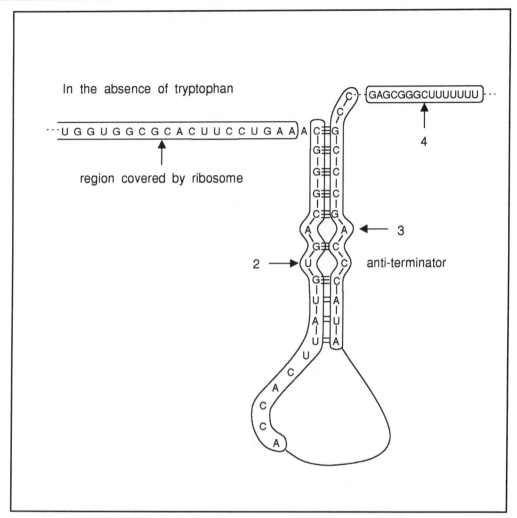

Figure 8.9 Structure of *trp* operon leader region transcripts in the absence of tryptophan. The whole of region 2 is available for binding to region 3, forming the anti-terminator (or pre-emptor). Region 4 is unpaired and hence not recognised by RNA polymerase as part of a terminator signal. Redrawn from B Lewin (1990), Oxford University Press.

In the absence of large amounts of tryptophan we have already seen that translation of the leader peptide stalls at the point where two tryptophan residues are required in a row. As the ribosome pauses before the stop codon, the region of RNA that it covers only extends to about nucleotide 72. The polymerase carries on with transcription and this time the whole of region 2 is available for formation of a stem-loop structure. As region 1 is unavailable for base paring due to ribosome attachment, region 2 interacts with region 3 as soon as it is produced by polymerase, forming the anti-terminator. Thus when the polymerase transcribes region 4, this cannot interact to form the terminator, and the enzyme continues transcription through the end of the leader sequence into the structural genes, resulting in operon expression.

SAQ 8.2	Try to decide, explaining your reasoning, what approximate level of expression of the *trp* operon there would be for the following.
	
	1) Normally (ie considering control of expression by repression and attenuation).
	
	2) If control occurred solely by repression, in the following *E.coli* strains under each of the following conditions:
	
	   a) wild type in the presence of low levels of tryptophan;
	
	   b) wild type in the presence of excess tryptophan;
	
	   c) *trp*L⁻ mutant in the presence of low levels of tryptophan;
	
	   d) *trp*L⁻ mutant in the presence of excess tryptophan;
	
	   e) O⁻ mutant in the presence of low levels of tryptophan;
	
	   f) O⁻ mutant in the presence of excess tryptophan.

8.5 Other operons controlled by attenuation

As mentioned above, many operons encoding amino acid biosynthetic enzymes are controlled by attenuation. Included are the *leu* (leucine), *thr* (threonine), *phe* (phenylalanine), and *his* (histidine) biosynthetic operons. Table 8.1 indicates the peptides coded by leader sequences for various operons, reinforcing the principle that they contain multiple codons for the regulatory amino acids. Remember, as for the *trp* operon, it is the level of amino acyl-tRNA that is commonly the vital factor in attenuation. Some operons (eg *thr*, *ilv*) are controlled by more than one amino acid. This is because these are related end products of a common pathway eg synthesis of isoleucine requires threonine as a precursor. Hence low levels of threonine engender low levels of isoleucine, and both amino acids have a regulatory effect on threonine synthesis. The spacing of the codons for regulatory amino acids is such that stalling of ribosomes at any one of them disrupts the formation of a termination stem-loop structure.

∏ Why do you think the *his* operon needs 7 his codons (see Table 8.1) in its leader region compared to only 2 *trp* codons in the *trp* operon.

This is a bit tricky, but first of all consider the *trp* operon. This is controlled by two mechanisms - repression and attenuation, and hence the attenuation mechanism can afford to be a little lax. We saw the effect of this dual regulation in SAQ 8.2. The fact that the *his* operon requires six histidine codons indicates that in this case, attenuation must be the major (in fact the only) control mechanism, and hence needs to be quite stringent.

To reinforce the ideas which we have formulated on attenuation using the *trp* operon as a model, it will be useful here to consider one of the other operons mentioned in a little more detail. We will look at the *leu* operon of *Salmonella typhimurium*. This encodes enzymes involved in leucine biosynthesis and has a 160bp leader sequence, as shown in Figure 8.10. This sequence has an open reading frame coding for a leader peptide of 28 amino acids, four of which are leucine residues. In addition, there are 8 regions (some overlapping) able to participate in stem-loop formation. These are able to pair as follows:

- region 1 - region 5;

- region 2 - region 3;

- region 4 - region 7;

- region 6 - region 8.

Operon	Leader Sequence											
His	Met	Thr	Arg	Val	Gln	Phe	Lys	[His	His	His	His	His]
	[His	His]	Pro	Asp								
Phe	Met	Lys	His	Ile	Pro	[Phe	Phe	Phe]	Ala	[Phe	Phe	Phe]
	Thr	Phe	[Pro]									
Leu	Met	Ser	His	Ile	Val	Arg	Phe	Thr	Gly	[Leu	Leu	Leu]
	[Leu]	Asn	Ala	Phe	Ile	Val	Arg	Gly	Arg	Pro	Val	Gly
	Gly	Ile	Gln	His								
Thr	Met	Lys	Arg	[Ile]	Ser	[Thr	Thr	Ile	Thr	Thr	Thr	Ile]
	[Thr	Ile	Thr	Thr]	Gly	Asn	Gly	Ala	Gly			
Ilv	Met	Thr	Ala	[Leu	Leu]	Arg	[Val	Ile]	Ser	[Leu	Val	Val]
	[Ile]	Ser	[Val	Val	Val	Ile	Ile	Ile]	Pro	Pro	Cys	Gly
	Ala	Ala	[Leu]	Gly	Arg	Gly	Lys	Ala				
Trp	Met	Lys	Ala	Ile	Phe	Val	Leu	Lys	Gly	[Trp	Trp]	Arg
	Thr	Ser										

Table 8.1 Leader peptides of various amino acid biosynthetic operons. (Adapted from Genes IV by B Lewin, Oxford University Press, UK, 1990). *Ilv* = Isoleucine, leucine, valine operon.

Pair 4) form the termination loop, pair 3) the anti-termination loop, and 1) and 2) form structures to support termination following ribosomal dissociation from the transcript.

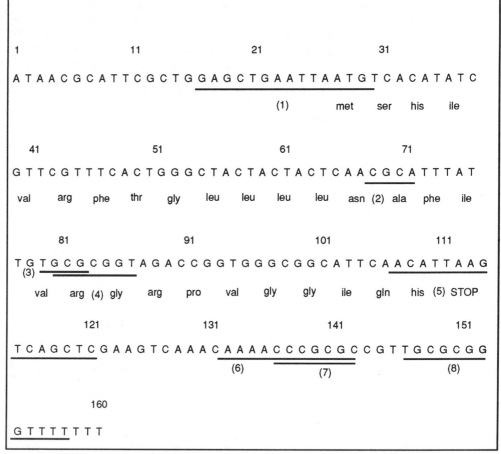

Figure 8.10 Leader sequence of the *S. typhimurium leu* operon. The sequence is that of the non-coding strand of the leader, ie that corresponding to the sequence of the RNA transcript, but having T instead of U residues, and is taken from Germill *et al* (1979) Proc Nat Acad Sci 76, 4941-4945.
The leader peptide sequence is shown below the corresponding nucleotide sequence. Underlined sequences are those involved in stem-loop structure formation, and are numbered for ease of reference.

SAQ 8.3

Using the information in Figures 8.9 and 8.10 and your knowledge of attenuation of the *trp* operon describe (with diagrams) what secondary structures are likely to form in the leader transcript in the presence of excess leucine and in its absence, and the consequences of these events for transcription of the structural genes in *Salmonella typhimurium*.

As we have seen, in the case of the amino acid biosynthetic operons considered so far, attenuation is governed by the availability of appropriate amino-acyl tRNA species. Other operons, however, make use of different signals to control attenuation, ie signals which do not involve leader sequence translation. For example, control of the *trp* operons of certain Gram-positive bacteria, although involving alternative leader RNA secondary structures, relies on the action of a specific regulatory molecule to determine between them. The regulatory molecule is the product of the *mtr* locus and, when activated by tryptophan, binds to the 5' region of the anti-terminator preventing its formation. Hence the terminator forms and transcription stops before the structural genes. The precise nature of the active species from *mtr* is uncertain - it may be a protein which is activated by tryptophan or an 'antisense RNA', which is an RNA sequence

antisense RNA

having the complementary sequence to the corresponding sense strand. In this case, the sense strand is the leader transcript.

Another mechanism of attenuation control is exercised by some of the pyrimidine biosynthetic operons of enteric bacteria. These use the level of UTP (a pyrimidine and an essential component of RNA) to control attenuation (eg the *pyr* Bi operon). This has a leader sequence encoding a 44 amino acid peptide. The leader transcript can form two stem-loop structures, a pause hairpin and a terminator, the former involving sequences closest to the promoter. The pause hairpin is followed by a series of U residues. These are spaced sufficiently far from the hairpin for it not to act as a terminator signal, but close enough that pausing of the RNA polymerase during transcription occurs for a relatively long time in the presence of limiting levels of UTP. In the presence of ample nucleotide the waiting time at the pause hairpin is relatively short, and the polymerase continues through the region of U residues to transcribe the terminator loop sequences.

Π   What effect do you think the increased pausing at the first hairpin in the absence of high levels of UTP may have on terminator loop formation?

It will prevent terminator formation. This occurs because whilst the polymerase pauses, ribosomes have time to bind to the transcript and move up to join the enzyme. Subsequently the two move together, and as soon as the first part of the terminator sequence is formed it is covered by a ribosome and hence unavailable for base-paring and terminator formation. In the presence of excess UTP, however, the polymerase pauses at the first hairpin for insufficient time to allow ribosomes to 'catch up' and hence continues to form a functional terminator.

You should have gathered by now that although a relatively straightforward idea, attenuation of each individual operon has been adapted for the most appropriate type of control, often reflecting the purpose of the operon.

In addition to those operons discussed, various others are controlled by cellular growth rate, and others by as yet undiscovered means.

SAQ 8.4	To clarify your thoughts on attenuation, list the important features of operons controlled in this way.

8.6 Control of gene expression by means other than repression/induction and attenuation

In this and the previous chapter, we have concentrated on control of transcription, and in particular on physical barriers to initiation and continuation of transcription. Another way in which transcriptional initiation may be governed is by the use of alternative sigma factors with the core RNA polymerase. It might be a good idea at this point for you to recap the role of sigma factor by reference to Chapter 5!

σ factor

The major sigma subunit of *E.coli* has a molecular weight of 70000 kD and is known as σ^{70}. Others also exist for use under specific conditions. The properties of these are summarised in Table 8.2 under normal conditions σ^{70} is used to activate RNA polymerase holenzyme. On being exposed to high temperature, however, the gene encoding σ^{32} is activated. The presence of a σ^{32} enables RNA polymerase to transcribe the 'heat shock' genes which produce a series of about 17 proteins in *E.coli* responsible for some sort of protective response(s), the precise nature of which is uncertain.

heat shock genes

∏ Examine Table 8.2 and try to decide how the different σ factors may be working.

common name	molecular mass (d)	-35 sequence	-10 sequence	distance between -35 and -10	genes transcribed
σ^{70}	70000	TTGACA	TATAAT	16-18bp	general
σ^{32} / σ^{H}	32000	CNCTTGAA	CCCCATNT	13-15bp	code for proteins produced in response to heat shock
σ^{54} / σ^{N}	54000	CTGGNA	TTGCA	6bp	code for proteins allowing cell to use alternatives to ammonia as nitrogen source.
σ^{28} / σ^{F}	28000	TAAA	GCCGATAA	15bp	those involved in chemotaxis and flagellar structure

Table 8.2 *E.coli* sigma factors. Data from B Lewin (1990), Genes IV, Oxford University Press, UK.

The important difference between the various factors is that they enable RNA polymerase to recognise different -35 and -10 promoter sequences, which are at various spacings relative to one another. This implies that it is the σ factor of the polymerase which causes it to contact the promoter at the -35 and -10 sequences. σ^{32} mRNA has a very short half life which is important in regulating rapid and short-lived heat shock responses. The σ^{32} gene (*rpo*H) is itself activated in response to heat by another minor σ factor, σ^{24}.

σ^{54} is normally present in small amounts in *E.coli*, and becomes active when the level of available ammonia in the medium drops. As shown in Table 8.2, it causes RNA polymerase to recognise and transcribe a variety of genes involved in utilisation of other nitrogen sources. σ^{28} is also present normally and allows transcription of genes causing bacteria make use of a variety of minor σ factors in similar ways to respond to different environmental conditions and to phage infection etc.

post-transcriptional regulation

In addition to control of gene expression by regulation of transcription, post-transcriptional regulation is also made use of in many cases. One of the various types of post-transcriptional control is the binding of small antisense RNAs to transcripts preventing their translation. This action may either be direct (ie the antisense RNA binds to a site on the transcript required for translational initiation) or indirect. In the case of indirect effects, the antisense RNA may bind to a region of the gene transcript such that the latter forms a revised secondary structure altering its activity. Control by antisense RNA molecules is absolute in that the molecule is either present (and

absolute control by antisense RNAs

therefore active) or absent. Hence the appearance and half-life of the regulator RNA must be strictly governed in terms of its initial transcription and its degradation by an appropriate ribonuclease. The best defined example to date of the regulatory action of an antisense RNA molecule on expression is in control of formation of the *omp*F gene product of *E.coli*. *Omp*F protein is an outer membrane protein of the bacterium, the synthesis of which is stopped when the osmolarity of the growth medium increases. The antisense RNA which prevents *omp*F synthesis comes from the *mic*F gene, and forms an RNA-RNA duplex with the *omp*F mRNA in the translation initiation region (see Figure 8.11).

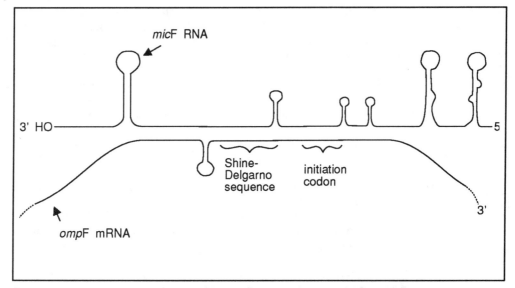

Figure 8.11 Inactivation of *omp*F mRNA by the *mic*F gene product. Adapted from Matthew & van Holde (1990), Biochemistry, Benjamin Cummings Publishing Co USA. Increased osmolarity in the medium activates a membrane-bound enzyme, *enz*V, which in turn activates a protein, *omp*R, which induces transcription of *mic*F. *mic*F RNA is partially complementary to the 5' end of *omp*F mRNA and binds to it forming a duplex such that *omp*F RNA cannot be translated.

A variety of other genes are regulated in ways similar to those which have been described in this and the previous chapter, and in other ways which there has not been room here to discuss. In particular, control at the level of translation and post-translational processing is an important topic to which we have not given much prominence with the exception of the brief description of control by antisense RNA. Post-translational processing is predominantly (but not the exclusive) the province of eukaryotic systems and is discussed in the BIOTOL text 'Genome Management in Eukaryotes'.

You should also realise that because DNA is more closely associated with proteins (histones) and carefully packaged into chromosomes, accessibility of the transcription machinery to DNA is especially a problem in eukaryotes.

Summary and objectives

The main focus of this chapter has been control of transcription by premature termination or attenuation. As we have seen this involves the adoption of differing secondary structures within the nascent RNA depending on the interaction and movement of ribosomes along a leader sequence within the message. The leader sequences of operons subject to control by attenuation lie between their promoters and structural genes.

Now that you have completed this chapter you should be able to:

- predict the likely trigger for the attenuation of a variety of operons;

- draw an overall structure for the *trp* operon showing the order of the major regulatory and structural genes;

- compare the ways in which repressor proteins controlling the *lac* and *trp* operons work;

- explain how stem-loop structures in the *trp* operon leader region are modified in the presence and absence of tryptophan;

- predict the level of expression of the *trp* operon of various mutants in the presence and absence of tryptophan;

- use the sequence of nucleotides in the leader region of operons to predict likely secondary structures that may be formed;

- list the important features of operons controlled by attenuation;

- explain how sigma factors may alter the extent of gene expression;

- give examples of how gene expression may occur at the post-transcriptional stage.

Responses to SAQs

Responses to Chapter 1 SAQs

1.1 It is merely necessary to remove all references to deoxyribose, since the sugar in ribonucleosides and ribonucleotides is ribose, and replace the row for thymine with one for uracil (see Section 1.1.1):

base	ribonucleoside[2]	ribonucleotide (eg 5′-monophosphate)[1]
adenine	adenosine	adenosine 5′-phosphate (AMP)[2] (or adenylate)[3]
guanine	guanosine	guanosine 5′-phosphate (GMP) (or guanylate)[3]
cytosine	cytidine	cytidine 5′-phosphate (CMP) (or cytidylate)[3]
uracil	uridine	uridine 5′-phosphate (UMP) (or uridylate)[3]

Nomenclature of ribonucleosides and ribonucleotides;[1] ie the single phosphate group is carried on C-5′ of the ribose;[2] other esters must be specified eg 2′-ATP means a triphosphate group is carried on C-2′;[3] these compounds may also be named as the corresponding acids, which are ionised at pH7.

1.2 1)

organism	A	G	C	5-Mec	T	A/T	G/C
vaccinia virus	29.5	20.6	20.0	-	29.9	0.99	1.03
E. coli	25.7	24.2	24.6	-	25.5	1.01	0.98
yeast	31.7	18.3	17.4	-	32.6	0.97	1.05
Chlorella vulgaris	20.2	30.0	26.4	3.45	19.8	1.02	1.01
broad bean	29.7	20.6	14.9	5.2	29.6	1.00	1.02
herring	27.8	22.2	20.7	1.9	27.5	1.01	0.98
frog	26.3	23.5	21.8	2.0	26.4	0.99	0.99
chicken	28.9	23.7	20.3	0.91	26.2	1.10	1.12
human	30.3	19.5	19.9	-	30.3	1.00	0.98

Therefore the data, within experimental error, do support Chargaff's ratios.

2) If A:T and G:C both approximate to 1.0, then G+A:C+T ie the Pu:Py will also be about 1:1.

1.3   1) a) If one strand of DNA contains 18% A, then for that strand C+G+T = 82%.

       b) The opposite (complementary) strand will contain 18% T (the complement of A) and A+C+G will = 82%.

    2) The complementary strand will be:

       3′ ...TTAAACGGCCTATCCGGGTA... 5′.

1.4   All that may be concluded is that C+G+T = 100 - 24.6 = 75.4%, since complementary basepairing does not occur.

1.5   Your completed table should look like this:

	A	B	Z
handedness	right	right	left
number of bases/turn	11	10	12
rotation/residue (°)	33	36	-30
rise/residue (nm)	0.255	0.34	0.37
pitch (nm)	2.8	3.4	4.5
diameter (nm)	2.55	2.37	1.84

If you have managed all this you are doing well! Any problems, then go over text again. All the answeres are there (somewhere!). You may find this table a useful summary of the properties of A, B and Z-DNA.

1.6   1) The sequence contains the inverted repeat (boxed):

       ...T GACCGA A T T C C TCGGTC G...
       ...A CTGGCT T A A G G AGCCAG C...

    this can form the cruciform:

    2) Since the double helical structure is stabilised by more hydrogen bonds than the cruciform, it is the more stable of the two.

1.7   DNA is composed of the sugar **deoxyribose**, the bases **adenine, guanine, cytosine** and **thymine** and **phosphate** groups. The bases of nucleic acids are classified into one of two groups either **purines** or **pyrimidines**. A combination of base and sugar forms a **nucleoside**; a base-sugar-phosphate is a **nucleotide**. Nucleic acids consist of **nucleotides** linked by **phosphodiester** bonds.

Watson and Crick proposed a structure for DNA based on the **X-ray diffraction** data of Franklin and Wilkins. Their proposal was that DNA was **double helical** with a distance between bases of **0.34 nm** and a pitch of 3.4 nm. The proposed structure had an approximate diameter of **2.0 nm**. The two strands **are antiparallel**, one strand running in the $5' \rightarrow 3'$ direction and the other $3' \rightarrow 5'$.

DNA can assume one of a number of secondary structures. The most common is **B-DNA**, but other conformations such as **A-DNA** and **Z-DNA** exist. **A-DNA** is favoured at low relative humidity. **Z-DNA** is formed at high ambient salt concentrations and alternating **Pu-Py** sequences.

See Section 1.3 for details, if you could not answer this.

1.8   A satisfactory response would be a table of the form:

Groove	Base	Chemical group	Acceptor/donator
minor	C	keto(oxo) oxygen (O^2)	acceptor
	T	keto(oxo) oxygen (O^2)	acceptor
	A	ring nitrogen (N^3)	acceptor
	G	ring nitrogen (N^3)	acceptor
		amino group on C^2	donor
major	A	ring nitrogen (N^7)	acceptor
		amino group of C^6	donor
	C	amino group of C^4	donor
	G	ring nitrogen (N^7)	acceptor
		keto(oxo) oxygen (O^6)	acceptor
	T	keto(oxo) oxygen (O^4)	acceptor

This was quite a difficult question and do not be too disappointed if you did not get it all right. The key to being able to answer this is to look at the functional groups of the bases and see which could accept and which could donate hydrogens. You might go through the groups listed above and mark those which are also involved in 'internal' hydrogen bonding during basepairing.

1.9   Length of DNA = 1.1 mm = 1.1×10^3 μm;

length of bacterial cell = 1.5 μm;

then ratio = $1.1 \times 10^3 \div 1.5 = 735$.

Hence the need for careful packaging of DNA into the bacterial cell.

1.10 After one replication in $^{14}NH_4Cl$, each DNA molecule would contain one strand labelled with ^{14}N and one strand labelled with ^{15}N; a ratio of 1:1. After a second replication the ratio of ^{14}N to ^{15}N would be 3:1. All the new deoxyribonucleotides would be synthesised from $^{14}NH_4Cl$. We would anticipate that half the DNA would have a density of light DNA, and half would have a density between light and heavy DNA. We can represent this in the following way:

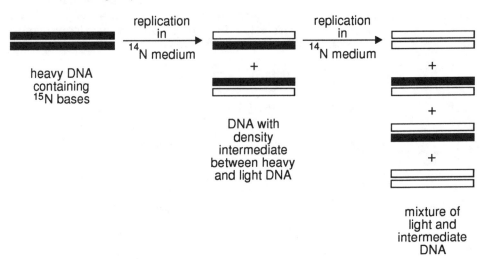

1.11 Length of *E. coli* chromosome = 1.1 mm = 1.1×10^6 nm. Therefore no. of bp = 1.1×10^6 / 0.34 therefore time for replication (given the presence of two replication forks) is $(1.1 \times 10^6/0.34) \div (800 \times 2 \times 60)$ minutes which is about 34 minutes.

1.12 Since the cell cycle lasts about half as long as the time to replicate the DNA, each *E coli* cell will contain about two chromosomal DNA molecules. In fact, analaysis of the cell contents of rapidly growing cultures shows an average of 2.3 DNA molecules per cell.

1.13 1) True, all viruses are obligate parasites.

2) False, some use a DNA-dependent RNA polymerase.

3) False, UTP not TTP. (TTP is of course used if DNA is an intermediate in the replication of viral RNA).

4) False, some make complementary RNA copies.

5) False, they may contain - or + copies, or sometimes both.

Responses to Chapter 2 SAQs

2.1   The genotype can be specified as *lac*⁻ and the phenotype as Lac⁻

2.2   1) Arg Ser Gly Asp.

2) a) Arg Gly Gly Gly;

b) Gln Asn Asn Asn.

2.3   1) The first two letters of the codon are the same in each case but the third letter can be A, C, U or G (see Table 2.1).

2) c) would be the most deleterious as it would lead to the greatest change in protein structure.

2.4   The number of mutants is $10^{12} \times 10^{-6} = 10^{6}$.

2.5   A single base change in DNA is called a **point mutation**. A **transition** is the change from one **pyrimidine** to another for example (C → T) or one **purine** to another (A → G). A **transversion** replaces a purine by a pyrimidine or vice versa. **Insertions** or **deletions** alter the reading frame of the gene and are examples of **frameshift** mutations.

Mutations can have a number of effects on the product of the gene. If the codon is altered but not the **amino acid** the mutation is a **silent** one.

A **mis-sense** mutation results in the incorporation of a **different** amino acid in the protein. If the mutated gene codes for an **essential** protein it may be a **lethal** mutation. **Conditional lethal** mutants contain, for instance, a heat sensitive enzyme that is inactivated at **35⁰C - 45⁰C**.

An **auxotroph** has one of the genes coding for the synthesis of an **essential** nutrient mutated. The non-mutant form is called the **wild-type** and will grow on a **minimal medium**.

2.6   1) Penicillin will kill the fast growing prototrophs.

2) Allows the growth of auxotrophs.

Mutant a is unable to synthesise riboflavin and biotin and has acquired two mutations. Note that this mutant requires both riboflavin and biotin to be present since there was no growth on plates 1, 7 or 11.

Mutant b cannot synthesise phenylalanine.

2.7   1) False. A minimal medium (not an enriched medium) containing penicillin to kill prototrophs is necessary.

2) True. Constitutive mutants arise from mutations which result in a gene being continuously expressed even in the absence of the inducer.

3) True.

2.8

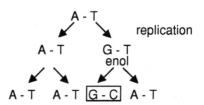

Note that in one of the products, the basepair A-T has been replaced by G-C.

2.9

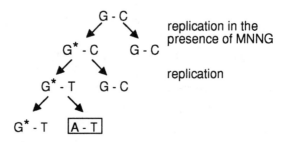

Note that G* preferentially pairs with T when it is replicated.

2.10 The % increase is = [(7-0.06)/0.06] x 100 = 11567%.

Responses to Chapter 3 SAQs

3.1   6) is correct. It can be calculated from 2n where n = number of divisions. (Thus $2^{21} = 2\ 097\ 152$).

3.2   The correct answer is 1). If you were incorrect you should read through the following definitions.

An auxotroph is a mutant micro-organism that cannot grow on minimal medium but requires the addition of some compound such as amino acid (eg proline or histidine) or a vitamin. A proline-requiring strain for example is referred to as *pro⁻*. A prototroph is a micro-organism which can grow on minimal medium. Streptomycin is a commonly used antibiotic with antibacterial properties. A sensitive organism is killed by streptomycin and is referred to as SmS.

3.3   The answer is 3) for the following reasons.

1) This is incorrect, because the initial bacteria strain is *leu⁺ his⁺* and thus the bacteria would be able to make their own leucine and histidine and hence would all grow on such a medium without DNA uptake.

2) This is incorrect, although bacteria which have taken up DNA will be selected for (only *leu⁺ his⁺* transformed bacteria will be able to grow) you would not be able to tell if the frequency of uptake of *leu⁺* was similar to that of *his⁺* or not, as we are using media lacking both additives we are only able to identify double transformed organisms.

3) Yes this is the correct answer. The frequencies of *leu⁺*, *his⁺* and *leu⁺ his⁺* transformants can be assessed and compared. If the frequencies of these were all similar then it is probable (but not proven) that genes encoding leucine and histidine synthesis are close together. It is however important to only use very small amounts of DNA to ensure that only one transformation event occurs per cell!

3.4

1	l
2	d
3	j
4	b
5	h
6	e
7	f
8	g
9	a
10	c
11	i
12	k

3.5   1) False. Sm prevents the Hfr from growing.

2) True.

3) True. The absence of the auxotrophically required amino acids stops the F from growing. The presence of the streptomycin stops the Hfr from growing.

4) True.

5) True. This marker is present after the shortest time (see also Figure 3.11).

6) False. They are *met+*, *leu+* and *trp+*.

7) True. Yes SmS is the marker which is effective against the donor.

3.6   This question is quite a puzzle. It is helpful to draw out a linear sequence for each Hfr. Thus:

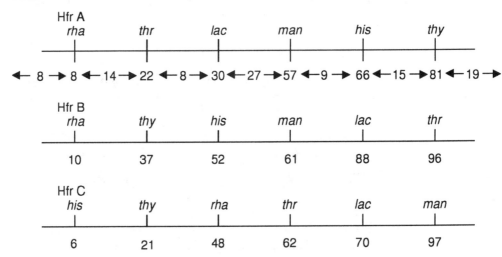

This immediately reveals that HfrA and HfrC insert genes in the same direction around the chromosome, but that the HfrB inserts them in the opposite direction.

We can also work out the time lag between each gene (shown for HfrA only).

We can therefore write the order of the genes around the circle as:

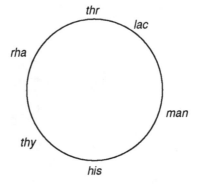

or more specifically we can impose on this order some more precise positions together with the insertion origins of the Hfr:

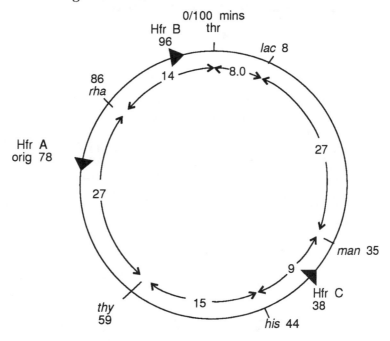

3.7

1) True, as the colonies can grow in the absence of d.

2) False, the recipient is *rec⁻* hence recombination of any incoming DNA into the recipient chromosome will not occur.

3) False, the Hfr is SmS.

4) True, this could be the case.

5) False. If the d⁺ gene is contained on an F′ plasmid, it may be maintained and expressed without integration into the F⁻ chromosome.

Responses to Chapter 4 SAQs

4.1 1) True.

2) False. This is almost true, but remember that the two ends (*cos*) are single-stranded.

3) False. This is false because the production and release of free phage is characteristic of the lytic not the lysogenic life-style.

4) False. The events described are in the wrong order. The order should be adsorption, followed by injection of the phage DNA into the bacterium and then its subsequent circularisation.

5) False. The attachment site on the host chromosome is *att* B whilst that on the tDNA is *att* P.

6) False. Site-specific is independent of *rec* A$^+$.

4.2 P'DEF *cos cos* ABCP.

4.3 1) True.

2) False. It is a repressor, blocking *oLpl* and *oRpR*.

3) True. It achieves this impact by destroying *c* I repressors.

4) False. The meaning of the word temperate in this context is that the phage can switch from lysogenic to lytic life-styles.

5) False. It also requires integrase.

6) True. If you thought this was false, re-read the last section again.

4.4 They are produced via aberrant excision from a normal prophage which itself does contain all the essential genes and hence gene products, therefore head and tail proteins are present in the cell to package the DNA. Only when the defective particle itself infects another cell is there a defiency in these products.

4.5 *xis* and *int* are required for the lysogenic life cycle and not the lytic life cyle and hence these λ*pbio* particles can form plaques.

4.6   The crossover event a) between homologous λ sequences that gives rise to λ^+/gal^-
recombinants and b) between homologous *gal* sequences that gives rise to λ^+/gal^+
recombinants.

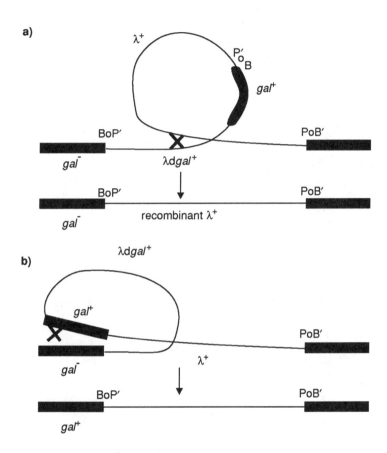

a)

b)

This is a difficult SAQ. If you have had difficulty with this SAQ it is worth persevering
and we suggest that you take another look at Sections 4.6 and 4.7.

4.7   Answer 1) is correct. We need to have excess bacteria with respect to phage to avoid
multiple infection of bacterial cells. Only the cells which have received his^+ genes will be
able to grow because of the absence of histidine in the medium. (Remember the recipient
cells are his^-. The absence of Ca^+ will prevent any wild-type (intact Pl) phage from
subsequently invading new hosts. We thus protect our newly transformed cells.

Answer 2) is wrong because we will get multiple infections. Thus bacteria which receive
his^+ genes are also likely to be infected by intact Pl cells and will be lysed.

Answer 3) is wrong because although we may successfully transform some bacterial
cells to his^+, intact Pl cells released from other cells will not be prevented from infecting
the his^+ cells.

We can represent this experiment in the following way:

in the presence of excess bacteria

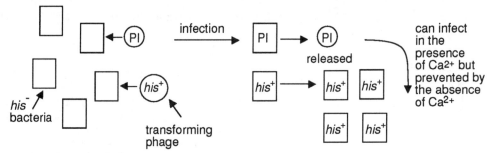

in the presence of excess phage

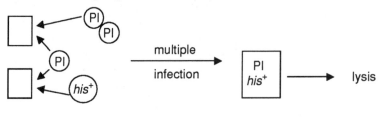

4.8

1) The answer is g) as this is the least frequent category of recombinant types.

2) From the fact that category g) is probably produced by a quadruple crossover we can tell that 1 is between 2 and 3, hence the gene order is: 2;1;3. We can represent the quadruple crossover in the following way:

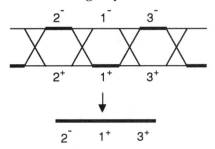

4.9

1) True.

2) True.

3) False. (See Figure 4.13).

4) False. No, kanomycin resistance is encoded by Tn5.

5) True.

6) True.

7) False. *tnpA* encodes for transposase.

Responses to Chapter 5 SAQs

5.1   Within the bounds of experimental error, the new RNA closely resembles the T2 phage DNA in its base composition, rather than the bacterial host DNA. The viral DNA must have acted as the template for the RNA. The synthesis of RNA follows the infection with DNA from the phage and precedes the manufacture of viral proteins. It is logical to assume that this RNA acts as an intermediary between the DNA code and the protein product. Note, though that the rRNA of *E. coli* does not directly reflect the total DNA base content. Those genes coding for rRNA are richer in some bases than the average sequence.

5.2   1) Each transcript requires a single RNA polymerase molecule. So at any one time there must be (approximately!);

800 + 100 + 700 = 1600 RNA polymerase molecules at work in transcription.

Many more RNA polymerase molecules are not involved actively in transcription at any one time.

2) The mRNA molecules, on average, have a much shorter half-life (time taken for half of the original mRNA to be degraded) than the rRNA and tRNA molecules. So a disproportionate amount of the cell's RNA polymerase (50%) is involved in transcribing mRNA.

5.3   1) Rate = 60 nucleotides per second; the time for 2000 nucleotides to be synthesised = 2000/60 = 33 secs.

2) The mRNA remains intact for a very short period of time especially when considering the time taken to manufacture the molecule. Almost as soon as the mRNA is made it starts to be degraded again. Hence, translation on the ribosomes must proceed quickly before the mRNA is lost. In fact, many of the ribosomes collected on an mRNA molecule for translation may appear to be physically attached to the bacterial chromosome.

5.4   There are small differences in the consensus sequences of A, B and C.

	-35 Box	-10 Box
a)	TTTACA	TATAAT
b)	TTGACA	TTAACT
c)	TTTACA	TATGAT

These small differences may lead to changes of promoter strength, but nevertheless the sequences are still very similar.

5.5   The correct order is: 3), 6), 2), 5), 1), 4).

Use the steps in SAQ 5.5 as labels for Figure 5.6.

5.6   We can see 4 ribosomes attached to the mRNA so this may represent the translation of 4 different polypeptides encoded by 4 different genes. However, each coding region represented on the mRNA may have more than one ribosome attached and translating it. A number of ribosomes translating a single mRNA is called a **polyribosome**. As this transcription/translation complex is dynamic we cannot confidently state the number of proteins produced from this message without further information.

At the 5′ end of the mRNA, an RNase is already degrading the mRNA which is still being transcribed (synthesised) at the 3′ end. The translation apparatus of the ribosome is physically linked to the DNA through the mRNA molecule.

5.7   1) Rifampicin is a potent inhibitor of chain initiation. Initiation occurs through the binding of the β subunit of RNA polymerase. This mutant is altered in the DNA sequence that codes for the β subunit, (see Section 5.2.3).

2) β subunit again! Streptolydigin inhibits chain elongation and presumably a different mutation in the same gene as in 1) has resulted in this phenotype. Together with 1) this demonstrates the β subunit's role in key catalytic functions; chain elongation and initiation, (see Section 5.5.3).

3) There are a number of possibilities including incorrect recognition of initation or termination signals or insertion of extra DNA into a structural gene. In this case, we actually have a processing mutant which is defective in the RNase E cleavage step for 5S RNA, (see Figure 5.8). This last example illustrates the fact that the phenotype of a mutant may not be conclusive and further experiments may be necessary. In this case, addition of normal RNase E to the abnormal 9S product would result in the normal product (5S RNA).

5.8   Your answer should look something like this:

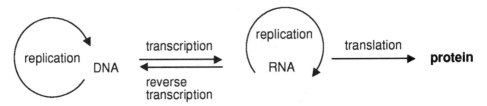

We must remember that the examples that we have discussed in these last sections are very much exceptions to the rule of one way flow of information.

If you find it difficult to apply these modifications, then check again with the previous sections on 'RNA as a template for RNA' and 'Reverse transcriptase' (Section 5.7), then compare them with the earlier section on 'The central dogma' (Section 5.1.2).

Responses to Chapter 6 SAQs

6.1 1) a) The code is based on triplets of mRNA bases.

b) The code is degenerate - most amino acids are coded by more than one synonym. This diminishes the effect of single base change mutations.

c) Two amino acids are specified by single codons - methionine by AUG and tryptophan by UGG.

d) There are three stop codons which do not code for amino acids but which cause termination of polypeptide synthesis. These are UAA, UGA and UAG.

e) The AUG codon is almost always used as the signal to initiate protein synthesis.

f) The code is almost universal. The exceptions are mainly changes in codon usage in certain mitochondrial genomes.

g) The code is unambiguous - each codon specifies either a specific amino acid or translational termination.

h) Codons with identical first and second bases always code for the same amino acid if the third base is a pyrimidine (U or C) and almost always do so if it is a purine (A or G).

2) The obvious experiment would be to produce a synthetic RNA containing the repeated sequence: ..CUCUCU... This molecule would then only contain two codons - CUC and UCU in alternate order. The first step would be to produce a DNA with a sequence as follows:

$$5' .....G\,T\,G\,T\,G\,T\,G\,T\,G\,T\,G\,T ..... 3'$$
$$.....C\,A\,C\,A\,C\,A\,C\,A\,C\,A\,C\,A.....$$

Transcription of this using RNA polymerase in the presence only of CTP and UTP will give the required RNA. If this is introduced into a cell-free protein synthesising extract containing amino acyl-tRNAs, all amino acids, amino acyl-tRNA synthetases, ribosomes etc, a protein containing the repeated motif shown below will be formed:

...Leu X Leu X Leu X...

where X is the amino acid specified by the codon UCU. (We already know that CUC specifies leucine). If a series of reactions are carried out with each tube containing a differently radioactively labelled amino acid, the amino acid encoded by UCU can be determined because only in the tube containing the labelled amino acid coded by UCU will a radioactively-labelled peptide have been formed.

6.2 1) Phenylalanine. This is coded by UU (C/U) ie two codons ending in either C or U at the 3' end. According to the wobble hypothesis these could both be recognised by a single tRNA with G as the 5' base of its anticodon.

2) Isoleucine. Coded by AUU, AUC and AUA. These could all be recognised by tRNA with inosine as the 5′ base of its anticodon (that is 5′-IAU-3′).

3) Serine. Coded by UCU, UCC, UCG, UCA, AGU and AGC. A minimum of three tRNAs would be required as shown below:

anticodon sequence	condons recognised
	UCU
5′-IGA-3′	UCC
	UCA
5′-CGA-3′	UCG
	AGU
5′-GCT-3′	AGC

4) Proline-coded by CCU, CCC, CCA and CCG. At least two tRNAs would be required to recognise these four codons - either with anticodons 5′-IGG-3′ and 5′-CGG-3′ or 5′-GGG-3′ and 5′-UGG-3′.

6.3 Amino acyl-tRNA anticodon-mRNA codon recognition involves only the triplets of bases, not the amino acyl group. One way of proving this might be to modify the side chain of the amino acyl residue attached to a particular tRNA to produce a different amino acid. By following incorporation of this amino acid into peptides produced from mRNAs of defined sequence, we can show that the tRNA will incorporate whichever amino acid is attached to it at positions governed by the codon with which the tRNA basepairs.

For example:

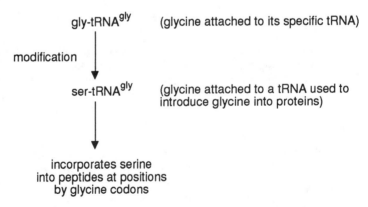

gly-tRNA^{gly} (glycine attached to its specific tRNA)

modification

ser-tRNA^{gly} (glycine attached to a tRNA used to introduce glycine into proteins)

incorporates serine
into peptides at positions
by glycine codons

6.4 Stage = initiation.

IF3. Prevents ribosomal subunit association prior to initiation of translation by binding to the 30S subunit.

IF1 and IF2. Prime the 30S subunit for mRNA interaction. When bound to GTP, IF2 enables fmet-tRNA$_f$ to bind to the initiation codon, producing the 30S initiation complex. They leave the complex on large subunit association and GTP hydrolysis.

Stage = elongation.

EF-Tu. Binds GTP, and in this form brings amino acyl-tRNAs to the ribosomal A site. It has endogenous GTPase activity, and only leaves the ribosome after GTP hydrolysis. This process is relatively slow to enable checking that the correct tRNA has bound. Peptide bond formation can only occur after EF-Tu-GDP has left the ribosome.

EF-Ts. Binds to EF-Tu-GDP to allow GDP-GTP exchange and reactivate Ef-Tu.

EF-G. Binds GTP, which it hydrolyses to provide free energy for translocation of the ribosome along the mRNA in the 5'-3' direction. This movement displaces uncharged tRNA and moves the growing peptidyl-tRNA to the P site.

Stage = termination.

RF1. Binds to stop codons, UAA and UAG, altering peptidyl transferase specificity such that the completed peptide chain is released by hydrolysis from its tRNA. Subsequently the ribosomal subunits dissociate, leaving the mRNA.

RF2. As RF1, but recognises stop codons, UAA and UGA.

Responses to Chapter 7 SAQs

7.1   a) Inducible negative control of gene expression (eg the *lac* operon):

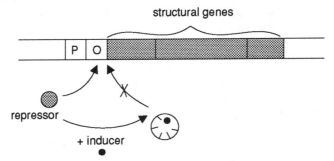

The repressor binds to the operator to prevent transcription, but on addition of the inducer, repressor is no longer able to bind to DNA. In other words, the inducer activates gene expression.

b) **Repressible negative control of gene expression (eg the *trp* operon):**

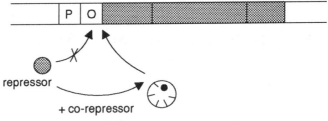

Repressor is inactive (and hence called an 'apo-repressor') in the absence of co-repressor ie co-repressor is required to inhibit gene expression.

c) **Inducible positive control of gene expression:**

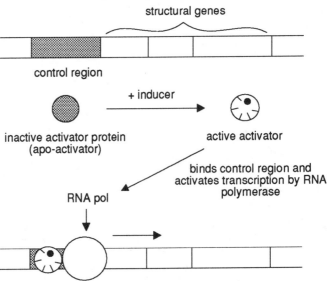

Transcription can only occur in the presence of activator protein at the control site of the operon. Activator can only bind on association with the inducer.

d) Repressible positive control of gene expression:

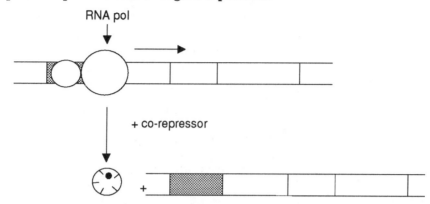

Activator protein is only able to enhance initiation of transcription in the absence of small modulator molecule. Binding of such a co-repressor causes dissociation of the protein from the operon and prevents gene expression.

The easiest way to remember these is firstly to understand the distinction between positive and negative control of expression. Positive control indicates that expression can only occur in the presence of an additional activator protein, whereas systems under negative control require the absence of active repressor protein for expression. The second thing is to learn what the effects of small modulator molecules on these processes are. These may be summarised in a small table, thus:

type of small molecule	effect on operon under positive control	effect of operon under negative control
inducer	enhances expression	enhances expression
co-repressor	reduces expression	reduces expression

7.2   The best way to start is to draw a sketch of the operon and *lac*I:

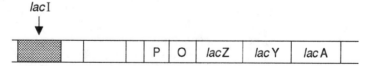

Strain A: normal *E. coli*. Without lactose: repressor binds operator and prevents transcription of structural genes - no expression. With lactose: the few remaining β-galactosidase molecules produce allolactose which acts as an inducer, binding to repressor and freeing the operator. RNA polymerase can now move through this region from the promoter to transcribe the structural genes.

Strain B: defective *lac*I gene ie produces no repressor: therefore the operon is expressed constitutively ie in the presence or absence of lactose.

Strain C: no functional operator ie cannot bind repressor protein. Thus we will have constitutive expression as in Strain B.

Strain D: no functional promoter ie cannot bind RNA polymerase. If RNA polymerase cannot bind to the operon there will be no expression under any conditions.

Strain E: defective *lac*Y gene ie no galactoside permease produced. The operon will never be active as no lactose will be able to enter the cell.

Strain F: defective *lac*Z gene ie no functional β -galactosidase produced. This strain will respond as Strain A in the absence of lactose, but expression of the operon will be much reduced in the presence of lactose. Lack of active β-galactosidase means that no allolactose will be produced to act as inducer. However, lactose itself can act as a weak inducer, so some expression will occur in its presence. But no metabolism of the lactose entering the cell will occur, and high levels may accumulate.

Strain G: defective *lac*A gene ie no functional galactoside transacetylase. This will have no effect on the response of the operon to lactose. The strain will react as Strain A. Any unusual galactosides, however, will not be acetylated and hence may accumulate in the cell.

7.3   This question tests your understanding of the cis- and trans-acting nature of the elements (control or structural genes) of the *lac* operon.

1) Strain B is *lac⁻* ie has no functional repressor gene. The repressor acts as a trans-acting factor, interacting with the operon operator. Introduction of a functional gene into the organism on a separate DNA strand will thus substitute for the defect. Responses to the presence and absence of lactose in this strain will be similar to those of Strain A above.

2) Strain C has a defective operator. This is a cis-acting control region ie the DNA sequence controls expression of the structural genes associated with it. These functions thus cannot be supplied 'in trans' (on a separate piece of DNA). This strain will show the same responsiveness to lactose as its parent, Strain C.

3) Strain D has a defect in the other cis-acting region, the promoter. As in 2) above, the presence of an intact promoter on a separate piece of DNA will not compensate for the defect. This strain will show the same *lac* operon expression as Strain D.

4) Strain E lacks a functional permease gene. Supplying a functional gene on a separate piece of DNA will compensate for this defect. The new gene will be constitutively expressed, producing permease which will allow lactose to enter the cell when available. The strain will act like wild type *E.coli* (Strain A) except that it will contain more permease molecules in the absence of lactose, and hence may respond more quickly in terms of switching the *lac* operon on when lactose become available. Compensation of the genomic defect by the introduced gene is possible because the importance of *lac*Y lies in its encoded protein.

5) Strain F lacks a functional *lac*Z gene and hence can make little response to the presence of lactose. Supplying a functional gene in trans will compensate for this defect, in a similar way to that described in 4).

6) The *lac*A gene product galactoside transacetylase, has little influence on control of the *lac* operon, as described in the answer to SAQ 7.2. The Strain G mutation will be compensated by introduction of functional *lac*A, and control of operon expression will occur as for Strain A.

7) Addition of a non-functional *lac*I gene will not influence expression in Strain A unless the defective gene encodes a mutant protein which competes with normal repressor and binds to the operator in such a way that it is not influenced by the allolactose inducer.

7.4

	ara	*lac*
inducer/activator	arabinose, cAMP	allolactose, (cAMP)
operator(s)	*ara* O_1, *ara* O_2, *ara* I	*lac* O
promoter(s)	C, BAD	ZYA main promoter, weaker upstream promoter
effect of absence of inducer	C protein binds to *ara* O_1 + *ara* O_2 to prevent *ara* C transcription. No binding to *ara* I	repressor binds to operator preventing transcription
effect of addition of inducer, low cAMP-CRP	C protein binds *ara* I to *ara* O_2 preventing transcription from *ara* BAD promoter	repressor leaves operator. RNA polymerase can transcribe ZYA but at very low efficiency. RNA polymerase also binds strongly to weaker promoter
effect of addition of inducer and cAMP-CRP	CRP binding causes C protein to release *ara* O_2 and activate RNA polymerase for *ara* BAD transcription	CRP binding enchances RNA polymerase transcription of the structural genes and prevents use of the weaker promoter

Responses to Chapter 8 SAQs

8.1   The *lac* operon is normally repressed when lactose levels are low. The protein binds to the operator preventing RNA polymerase from moving onto the structural genes. In the presence of lactose, the gene products of this operon are required and some of the first few molecules of lactose which enter the cell are changed into allolactose by the few residual molecules of β-galactosidase. This compound acts as an inducer, binds to the repressor and causes it to change conformation, leaving the operator. RNA polymerase bound to the promoter can now move onto the structural genes and transcribe them.

The *trp* operon, on the other hand, is active in the absence of cellular tryptophan. The repressor, *trp*R, can only bind to the operator and prevent transcription from starting in the presence of tryptophan as a co-repressor.

So, in summary, the catabolic *lac* operon is only active in the presence of its substrate, a form of which acts as an inducer, removing the repressor from the operator. The biosynthetic *trp* operon on the other hand is inactivated in the presence of its product (tryptophan) which acts as a co-repressor, activating the repressor protein and allowing its interaction with the operon to prevent transcription.

8.2

	Effect on Expression	
mutant and tryptophan level	**control by repression and attenuation (1)**	**control by repression only (2)**
wild-type - tryptophan	moderate to high level of expression, some repression due to low level of trp present, also small degree of attenuation by the tRNA^{trp}	high level of expression, only a small degree of repression and no attenuation
wild-type + tryptophan	minimal expression: maximal repression and maximal attenuation of any transcripts that escape repression	low expression, no attenuation, so although repression is maximal some expression occurs due to a small number of RNA polymerase molecules escaping repression
*trp*L⁻ - tryptophan	good expression as no leader sequences for any attenuation effect and no tryptophan to activate repressor	as in (1)
*trp*L⁻ + tryptophan	poor expression due to activation of repressor and binding to operator: no attenuation effect however	as in (1)
O⁻ - tryptophan	no repressor effect so maximal initiation of transcription: lack of trp causes ribosomal stalling and prevents termination loop formation leading to high expression	no functional operator for repressor binding. No attenuation mechanism hence maximal possible expression
O⁻ + tryptophan	no repressor effect due to non-functional operator, therefore maximal initiation of transcription: presence of excess trp allows termination loop formation hence, low-level expression, but not as low as if functional operator present also	as in (1)

8.3 Consult Figure 8.9

In the presence of excess leucine:

RNA polymerase will first pause on formation of a hairpin loop between regions 2 and 3. A ribosome binding upstream will move through onto the reading frame and catch up with the polymerase. The enzyme and ribosome will continue in tandem to the stop codon, and polymerase will proceed alone from there. Hence the only regions available initially for stem-loop formation will be regions 6 and 8, forming the attenuator or termination loop:

```
                    G — U
                 C         U
                   C = G
                   G = C
                   C = G
                   G = C
                   C = G
                   C = G
                   C = G
                   A = U
                   A = U
                   A = U
                   A = U
  5' ─────────────────/    \─ U U U ─────── 3'
```

When the ribosome leaves the transcript and before another one adjoins it, two support secondary structurs form by pairing of regions 1 and 5 and 2 and 3 as follows:

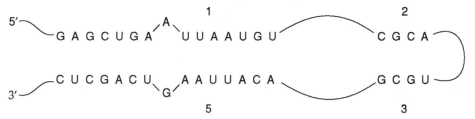

The effect will be inhibition of gene expression as transcription terminates before the structural genes.

In the absence of leucine:

RNA polymerase will proceed to the end of region 3 and formation of the region 2-3 stem loop will cause it to pause. Meanwhile a ribosome will bind and move along translating the leader region until it reaches the 4 leucine codons. At this point it stalls due to lack of tRNA^{leu} although it will have disrupted the region 2-3 RNA secondary structure, allowing RNA polymerase to continue. The RNA subsequently forms the next possible secondary structure - a stem loop structure due to basepairing between regions 4 and 7 as follows:

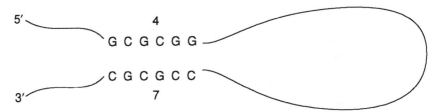

This forms the pre-emptor loop preventing formation of the attenuator/terminator loop by making region 6 unavailable for binding to region 8. Hence the effect is to allow RNA polymerase to continue and transcribe the structural genes.

8.4   Operons controlled in this way have the following characteristics:

1) They contain a leader region between the promoter and structural genes which is transcribed.

2) Sequences within the leader region are able to form alternative secondary structures to influence continuation of transcription by RNA polymerase. At least two alternative secondary structures must be possible and mutually exclusive - a pre-emptor and an attenuator structure.

3) The signals which influence formation of alternative secondary structures in leader RNA usually relate to the function of the operon under consideration.

4) Operons may be controlled by attenuation alone, in which case the signal regions within the leader must be unambiguous and guaranteed to be noted, or by attenuation in conjunction with other mechanisms. Such operons show a wider range of levels of gene expression than others.

Appendix 1

Units of measurement

For historical reasons a number of different units of measurement have evolved. The literature reflects these different systems. In the 1960s many international scientific bodies recommended the standardisation of names and symbols and a universally accepted set of units. These units, SI units (Systeme Internationale de Unites) were based on the definition of: metre (m), kilogram (kg); second (s); ampare (A); mole (mol) and candela (cd). Although, in the intervening period, these units have been widely adopted, their adoption has not been universal. This is especially true in the biological sciences.

It is, therefore, necessary to know both the SI units and the older systems and to be able to interconvert between both sets.

The BIOTOL series of texts predominantly uses SI units. However, in areas of activity where their use is not common, other units have been used. Tables 1 and 2 below provides some alternative methods of expressing various physical quantities. Table 3 provides prefixes which are commonly used.

Mass (SI unit: kg)	Length (SI unit: m)	Volume (SI unit: m^3)	Energy (SI unit: $J = kg\ m^2\ s^{-2}$)
$g = 10^{-3}\,kg$	$cm = 10^{-2}\,m$	$l = dm^3 = 10^{-3}\,m^3$	$cal = 4.184\ J$
$mg = 10^{-3}\,g = 10^{-6}\,kg$	$Å = 10^{-10}\,m$	$dl = 100\ ml = 100\ cm^3$	$erg = 10^{-7}\ J$
$\mu g = 10^{-6}\,g = 10^{-9}\,kg$	$nm = 10^{-9}\,m = 10Å$	$ml = cm^3 = 10^{-6}\,m^3$	$eV = 1.602 \times 10^{-19}\ J$
	$pm = 10^{-12}\,m = 10^{-2}\,Å$	$\mu l = 10^{-3}\,cm^3$	

Table 1 Units for physical quantities

Concentration (SI units: $mol\ m^{-3}$)

a) $M = mol\ l^{-1} = mol\ dm^{-3} = 10^3\ mol\ m^{-3}$

b) $mg\,l^{-1} = \mu g\ cm^{-3} = ppm = 10^{-3}\ g\ dm^{-3}$

c) $\mu g\ g^{-1} = ppm = 10^{-6}\ g\ g^{-1}$

d) $ng\ cm^{-3} = 10^{-6}\ g\ dm^{-3}$

e) $ng\ dm^{-3} = pg\ cm^{-3}$

f) $pg\ g^{-1} = ppb = 10^{-12}\ g\ g^{-1}$

g) $mg\% = 10^{-2}\ g\ dm^{-3}$

h) $\mu g\% = 10^{-5}\ g\ dm^{-3}$

Table 2 Units for concentration

Fraction	Prefix	Symbol	Multiple	Prefix	Symbol
10^{-1}	deci	d	10	deka	da
10^{-2}	centi	c	10^2	hecto	h
10^{-3}	milli	m	10^3	kilo	k
10^{-6}	micro	μ	10^6	mega	M
10^{-9}	nano	n	10^9	giga	G
10^{-12}	pico	p	10^{12}	tera	T
10^{-15}	femto	f	10^{15}	peta	P
10^{-18}	atto	a	10^{18}	exa	E

Table 3 Prefixes for S1 units

Appendix 2

Chemical Nomenclature

Chemical nomenclature is quite a difficult issue especially in dealing with the complex chemicals of biological systems. To rigidly adhere to a strict systematic naming of compounds such as that of the International Union of Pure and Applied Chemistry (IUPAC) would lead to a cumbersome and overly complex text. BIOTOL has adopted a pragmatic approach by predominantly using the names or acronyms of chemicals most widely used in biologically-based activities. It is recognised however that there remains some potential for confusion amongst readers of different background. For example the simple structure CH_3COOH can be described as ethanoic acid or acetic acid depending on the environment or industry in which the compound is produced or used. To reduce such confusion, the BIOTOL series makes every effort to provide synonyms for compounds when they are first mentioned and to provide chemical structures where clarity and context demand.

Appendix 3

Abbreviations used for the common amino acids

Amino acid	Three-letter abbreviation	One-letter symbol
Alanine	Ala	A
Arginine	Arg	R
Asparagine	Asn	N
Aspartic acid	Asp	D
Asparagine or aspartic acid	Asx	B
Cysteine	Cys	C
Glutamine	Gln	Q
Glutamic acid	Glu	E
Glutamine or glutamic acid	Glx	Z
Glycine	Gly	G
Histidine	His	H
Isoleucine	Ile	I
Leucine	Leu	L
Lsyine	Lys	K
Methionine	Met	M
Phenylalanine	Phe	F
Proline	Pro	P
Serine	Ser	S
Threonine	Thr	T
Tryptophan	Trp	W
Tyrosine	Tyr	Y
Valine	Val	V

Appendix 4

Abbreviations and nomenclature of nucleic acids and derivatives

The abbreviations employed in this book are based on those proposed by the Commission on Biochemical Nomenclature (CBN) of the International Union of Pure and Applied Chemistry (IUPAC) and the International Union of Biochemistry (IUB).

Nucleosides

A	adenosine
G	guanosine
C	cytidine
U	uridine
ψ	5-ribosyluracil (pseudouridine)
I	inosine
X	xanthine
rT	ribosylthymine (ribothymidine)
N	unspecified nucleoside
R	unspecified purine nucleoside
Y	unspecified pyrimidine nucleoside
dA	2'-deoxyribosyladenine
dG	2'-deoxyribosylguanine
dC	2'-deoxyribosylcytosine
dT or T	2'-deoxyribosylthymine (thymidine)

Nucleotides

AMP	adenosine 5'-monophosphate
GMP	guanosine 5'-monophosphate
CMP	cytidine 5'-monophosphate
UMP	uridine 5'-monophosphate
dAMP	2'-deoxyribosyladenine 5'-monophosphate
dGMP	2'-deoxyribosylguanine 5'-monophosphate
dCMP	2'-deoxyribosylcytosine 5'-monophosphate
dTMP	2'-deoxyribosylthymine 5'-monophosphate

2'-AMP, 3'-AMP, 5'-AMP etc	2'-, 3'-, 5'-phosphates of adenosine etc
ADP etc	5'-(pyro)diphosphates of adenosine etc
ATP etc	5'-(pyro)triphosphates of adenosine etc
ddTTP etc	2'-, 3'- dideoxyribosylthymine 5'-triphosphate
araCTP	1-β-D-arabinofuranosylcytosine 5'-triphosphate

Polynucleotides

DNA	deoxyribonucleic acid
cDNA	complimentary DNA (or copy DNA)
mtDNA	mitochondria DNA
RNA	ribonucleic acid
mRNA	messenger RNA
rRNA	ribosomal RNA
tRNA	transfer RNA
nRNA	nuclear RNA
hnRNA	heterogenous nuclear RNA
snRNA	small nuclear RNA
Alanine tRNA or tRNA^{Ala}	transfer RNA that normally accepts alanine
Ala-tRNA^{Ala} or Ala-tRNA	transfer RNA that normally accepts alanine with alanine residue covalently linked

Miscellaneous

RNase, DNase	ribonuclease, deoxyribonuclease
Pi, PPi	inorganic orthophosphate and pyrophosphate
nt	nucleotide
bp	basepair
mt	mitochondrial
cp	chloroplast

Index